Series

Work Out

Biology

'A' Level

The titles in this series

MACMILLAN
MASTER
SERIES

Work Out

Biology

'A' Level

G. W. Stout
and
N. P. O. Green

MACMILLAN

First published 1986

Published by
MACMILLAN EDUCATION LTD
Houndmills, Basingstoke, Hampshire RG21 2XS
and London
Companies and representatives
throughout the world

Typeset by TecSet Ltd
Printed by The Bath Press, Avon

British Library Cataloguing in Publication Data
Stout, G.W.
Work out biology : 'A' Level.—(Macmillan work out series)
1. Biology—Examinations, questions, etc.
I. Title   II. Green, N.P.O.
574′.076   QH316
ISBN 0-333-39184-5

Contents

Preface

The aim of this book is to help you to prepare for your 'A' level Biology examinations. In doing this, it differs from most other study guides and revision guides by giving you actual answers to questions rather than guidance as to how to approach the questions.

It has been the authors' experience as teachers and examiners that what examination candidates really need to see are actual answers. These answers are not to be taken as 'definitive' answers. They are not necessarily the only answers which could satisfactorily answer the questions. To this extent they are not to be considered as 'model' answers. At 'A' level it is doubtful if there is such a thing.

Cambridge and Horsham, 1986

G.W.S.
N.P.O.G.

Acknowledgements

Figure 6.1 by the late K. Murray is reproduced by kind permission of his wife.

The cover photograph, by Manfed Danegger NHPA, shows *Anas platyrhynchos* (mallard) taking off from a lake.

The authors and publishers also wish to thank the following who have kindly given permission for the use of copyright material:

The Associated Examining Board, the Joint Matriculation Board, the Northern Ireland Schools Examination Council, the Oxford and Cambridge Schools Examination Board, the Scottish Examination Board, the University of Cambridge Local Examinations Syndicate, the University of London Schools Examinations Board and the University of Oxford Delegacy of Local Examinations for questions from past examination papers.

Every effort has been made to trace all the copyright holders but if any have been inadvertently overlooked the publishers will be pleased to make the necessary arrangement at the first opportunity.

The University of London Entrance and School Examinations Council accepts no responsibility whatsoever for the accuracy or method in the answers given in this book to actual questions set by the London Board.

Acknowledgement is made to the Southern Universities' Joint Board for School Examinations for permission to use questions taken from their past papers but the Board is in no way responsible for answers that may be provided and they are solely the responsibility of the authors.

The Associated Examining Board, the University of Oxford Delegacy of Local Examinations, the Northern Ireland Schools Examination Council, the University of Cambridge Local Examinations Syndicate and the Scottish Examination Board wish to point out that worked examples included in the text are entirely the responsibility of the authors and have been neither provided nor approved by the Board.

Examination Boards for Advanced Level

Syllabuses and past examination papers can be obtained from:

The Associated Examining Board (AEB)
Stag Hill House
Guildford
Surrey GU2 5XJ

University of Cambridge Local Examination Syndicate (UCLES)
Syndicate Buildings
1 Hills Road
Cambridge CB1 2EU

Joint Matriculation Board (JMB)
78 Park Road
Altrincham
Cheshire WA 14 5QQ

University of London School Examinations Board (L)
University of London Publications Office
52 Gordon Square
London WClE 6EE

University of Oxford
Delegacy of Local Examinations (OLE)
Ewert Place
Summertown
Oxford OX2 7BZ

Oxford and Cambridge Schools
Examination Board (O & C)
10 Trumpington Street
Cambridge CB2 1QB

Scottish Examination Board (SEB)
Robert Gibson & Sons (Glasgow) Ltd
17 Fitzroy Place
Glasgow G3 7SF

Southern Universities, Joint Board
(SUJB)
Cotham Road
Bristol BS6 6DD

Welsh Joint Education Committee
(WJEC)
245 Western Avenue
Cardiff CF5 2YX

Northern Ireland Schools Examination
Council (NISEC)
Examinations Office
Beechill House
Beechill Road
Belfast BT8 4RS

1 Introduction

1.1 How to Use This Book

Each chapter in this book has been written to provide you with the following:

1. Clear **definitions** of important terms used in biology.
2. A short **summary** of the main facts, terms and principles associated with each major topic area. The topic areas are listed as sections, e.g. 6.2.
3. **Answers** to a selection of past examination questions from a number of examination boards and some prepared specially for this book. These questions have been selected to reflect the frequency with which topics tend to be tested in examinations and to illustrate certain topics which are known to cause candidates problems.

It is suggested that you use this book throughout your 'A' level course when preparing answers for homework and when revising for end of term and 'mock' examinations. This will give you familiarity with the book and develop good practice in examination techniques. It is assumed that you will have access to a 'sound' textbook* throughout your course and that this will provide you with the detail which a revision text cannot supply.

In some types of question where 'one-word' answers are required there is usually only one correct answer, but there may be many ways of expressing this, e.g. fermentation/anaerobic respiration/anaerobiosis. Any of these terms would probably be marked correct. Likewise, in essay-type answers there are many ways of presenting acceptable answers and many candidates can achieve the same overall mark by many different answers. You should bear this in mind when studying the answers given in this book.

The purpose in providing many of the answers in this book is to show you the breadth and depth of knowledge required in answers. In some cases you may be surprised at either how much detail or how little detail is required. It is worth taking careful note of this as it will help you when preparing your own answers.

1.2 Examinations and Assessment

The purpose of examinations is to measure attainment in a certain subject. You will be assessed against two criteria, namely *absolute* standards linked to the skills, abilities and concepts associated with the subject and *relative* standards of your fellow students. The term criterion-referenced applies to the former case and norm-referenced applies to the latter. 'A' level assessment is a blend of both forms of assessment.

All examination syllabuses list the abilities to be tested by the examination. These skills and abilities are related in a broad way to the individual's 'biological intelligence'. Examiners aim to test all of these skills and abilities in their

*For example: N.P.O. Green, G.W. Stout and D.J. Taylor, *Biological Science*, vols. 1 and 2, ed. R. Soper, published by Cambridge University Press (1984, 1985).

examination papers. The skills and abilities are hierarchical, which means that a question which test skills at Level 5 will also test skills at all lower levels. For theory papers these may be summarized as follows:

1. Recall of **knowledge** of learned facts, terms and principles (sometimes called 'specifics' and 'processes').
2. **Comprehension** of these facts, terms and principles presented in a variety of ways, e.g. verbal, diagrammatic, graphical and tabular, and the translation and interpretation of this information.
3. **Application** of the data described in 2 above to new situations.
4. **Synthesis** and **analysis** of data in terms of the construction of hypotheses and the design of experiments.
5. **Evaluation** of biological information so as to display mastery of data at all levels as presented above.

Underlying all the above abilities you must also be able to communicate knowledge relevantly, logically and clearly.

In practical examinations the skills and abilities to be tested may be summarized as the ability to:

1. observe accurately,
2. record observations precisely,
3. follow instructions,
4. manipulate familiar and unfamiliar biological materials,
5. interpret data,
6. pursue logical developments.

The above skills and abilities may be tested in a variety of ways which may involve the use of living and dead organisms, the microscope, dissection, physiological and behavioural experiments and fieldwork.

1.3   Guide to Study and Revision

You will need to work steadily and conscientiously *throughout* your A-level course. Last-minute revision is useful but cannot compensate for poor work earlier in the course.

Too many students approach 'A' level work with the idea that it is simply a continuation of 'O' level. This is a dangerous and false approach. The intellectual and academic demands of 'A' level are far greater than 'O' level. For students with this attitude the 'A' level course will prove to be difficult and frustrating as it will be beyond their capability. These are the students who will eventually gain grades O and F.

At 'A' level it is important that you appreciate that you will need to assimilate a substantial amount of subject specific material, e.g. topics such as biochemistry, physiology, genetics etc. and be able to manipulate this material in a variety of different ways, e.g. involving skills and abilities such as knowledge, comprehension, application etc. At 'O' level most students rely heavily on their teachers, a single textbook and guided revision lessons. At 'A' level this is not the case. To perform well it is therefore important that you are able to undertake responsibility for your own work. This means that you must organize your own study throughout the course and assume responsibility for your own revision.

It is important that you develop a genuine and deep interest in the subject as this will provide the motivation and self-discipline needed to sustain you through the course. Reading 'around' the subject is important. You should follow up topics covered in lessons by looking them up in more specialist books. It is also worth reading journals and periodicals such as *New Scientist*, *Scientific American* and the *Journal of Biological Education*. Most schools will have copies of these; if not, your local library may be able to help. Above all, you must work methodically and with determination. Many potentially good students at 'A' level fail to obtain the grades they should aspire to, simply because they neglect the advice given above. They rely too much on last-minute cramming and they pay the awful price for this.

It should not be thought, though, that final revision is unnecessary. You can organize your final revision in such a way that it becomes enjoyable and highly productive. The following advice should help you with this.

1. Obtain an examination syllabus and recent past examination papers. These can usually be purchased from the examination boards whose addresses are given on pages ix and x.
2. Ensure that you are familiar with the form of each paper and the style of questions.
3. Ensure that all your course notes are complete. All the information that you require for your revision should be in your notebooks. Avoid revising from textbooks. Use them only for reference purposes.
4. Work out a programme of revision and keep to it. This discipline is essential for revision.
5. Revise the course topic by topic. The chapter headings and their order given in the contents list of this book may be helpful.
6. Read through your notes on a particular topic and make a list of sub-topics, e.g.

Topic   *Human alimentary canal*
Sub-topics *buccal cavity*
                teeth
                oesophagus
                peristalsis
                stomach
                duodenum
                pancreas
                liver
                ileum
                etc.

Study each of the sub-topics in detail. Begin by reading your notes on this several times.

(a) Make a list of words that need definitions, e.g. ingestion, digestion, and learn how to use these terms correctly.
(b) List the key facts associated with the structure or function and ensure that you understand the implications of all these biological terms and definitions.
(c) Identify any principles or concepts which appear, and which can be applied to other situations, e.g. diffusion, surface area/volume ratio, counter current exchange.

7. Tables of information should be thoroughly studied and the details learned. A helpful way of memorizing tabulated information is to use a mnemonic.

One way of doing this is to arrange the words so that their initial letters spell out a word, e.g. the characteristics of all living organisms are:

movement, **a**ssimilation, **r**espiration, **r**eproduction, **i**rritability, **g**rowth, **e**xcretion.

This spells out (phonetically) marri(a)ge.

8. Diagrams can only be learned by drawing them several times and looking for patterns in the way they are drawn. For example, the structure of the eye appears to be based on concentric circles which correspond to layers and regions, e.g. sclerotic, choroid and retina. Simplify the diagrams that you draw so that they can be drawn and labelled in about 3 to 4 minutes (see notes on diagrams on page 11).

9. Many complex biological activities which involve many structures and functions can be simplified into the form of a flow diagram. Constructing a flow diagram is a valuable way of revising as it helps you to appreciate the interrelationships between structures and functions. It also encourages you to think logically and record information concisely. Answering essay questions certainly requires these skills.

To construct a flow diagram you take a sheet of paper and write down the main structures and/or features associated with the topic being studied. Use each one of these words as a stimulus to trigger other associated words. The various words can be linked with lines and arrows to indicate functional relationships.

In the example below (see Fig. 1.1) dealing with gaseous exchange in mammals, the three original words were:

atmosphere: the source of oxygen
lungs: the gaseous exchange organ
tissues: the source of carbon dioxide.

Other words were added, where appropriate, using the three words above as stimuli, to produce Fig. 1.1.

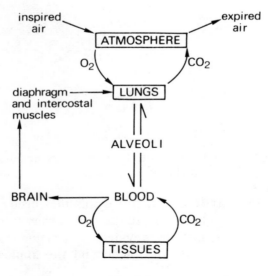

Fig. 1.1

Finally, a complete flow diagram can be built up as shown in Fig. 1.2.

10. Following your revision of each topic, read the appropriate chapter in this book. Note the **definitions** and **key words**. It is vital that you should know

4

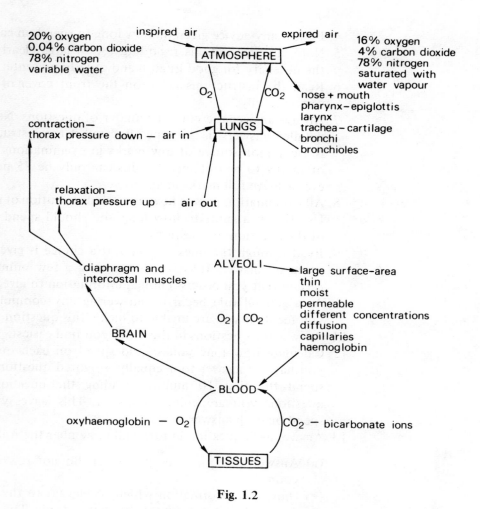

inspired air — expired air

ATMOSPHERE

20% oxygen
0.04% carbon dioxide
78% nitrogen
variable water

16% oxygen
4% carbon dioxide
78% nitrogen
saturated with
water vapour

O_2 — CO_2

LUNGS

nose + mouth
pharynx – epiglottis
larynx
trachea – cartilage
bronchi
bronchioles

contraction—
thorax pressure down — air in

relaxation—
thorax pressure up — air out

diaphragm and
intercostal muscles

ALVEOLI

large surface-area
thin
moist
permeable
different concentrations
diffusion
capillaries
haemoglobin

BRAIN

O_2 CO_2

BLOOD

oxyhaemoglobin — O_2 CO_2 — bicarbonate ions

TISSUES

Fig. 1.2

what each of the key word structures looks like, where it is found, and how it functions. Likewise, each key word process should be clearly understood in terms of where it occurs, how it occurs, and why it occurs. Concepts and principles too should be identified as they appear in your revision.

Read the questions and answers given in each chapter. The answers are given as a guide to content and style. Many of these questions can be answered correctly in a number of ways and the answer given here may not be the *only* answer.

Finally, try answering some of the questions from past papers. Do not aim at speed at this stage. Aim at answering the questions correctly. If, at a later stage, you find you are still taking a very long time to produce an answer, particularly an essay-type answer, you are writing too much. Remember, your answers must be relevant, accurate, logically developed and clearly expressed. If you satisfy these four criteria, success will be yours.

1.4 Examination Technique

1. Fill in your name, centre number and candidate number as instructed.
2. Note the number of questions that have to be answered and any special requirements, such as, 'Answer **one** question from Section B and **one** question from Section C.'
3. Answer **all** compulsory questions.

4. Follow any advice given on how long to spend on each section.

5. Note any instructions regarding use of large clearly labelled diagrams and the necessity for good English and orderly presentation in your answers.

6. Read the instructions given on the front cover of the examination paper very carefully.

7. Always answer the correct number of questions. Never decide to give three detailed answers in the time allocated if the instructions ask for four. This is a common source of low marks in examinations. The maximum number of marks to be obtained by this can only be 75 per cent, and few people ever achieve full marks in an answer.

8. All examination papers now give the distribution of marks for each question. Use this as a guide to how long you should spend on answering each part of the question or questions.

9. Read through the questions and, if a choice is given, decide which you are going to answer. It is worth spending a few minutes over this and being certain that you possess enough information to give complete answers.

10. As a general rule begin by answering any compulsory question(s), unless you feel that you are unable to tackle this question well.

11. Answer the questions in the order you find easiest.

12. Calculate how long you should spend on each question. For example, if you have to answer four equally weighted questions in $2\frac{1}{2}$ hours, you may spend the first 10 minutes reading the questions and deciding which questions you are going to answer. This leaves you with 35 minutes to spend on each answer.

13. Answer each question in turn and remember the following:

    (a) Answer the question that is set; do not rewrite the question to suit yourself.
    (b) Only give information which is relevant to the question. Do not write down all that you know about that topic. This will not gain marks and will waste your time.
    (c) Answers should be concise and written in good English. Generally this is best achieved by writing short sentences.
    (d) Present your answer in a logically developed way.

14. Remember to stick to your time allocation. After writing, for, say, 35 minutes stop, and begin the next question. Do this even if you have not completed all the question.

15. Remember that the examiner is looking for the correct use of biological facts in your answer. Marks can only be awarded for what you write on the paper—not for what remains in your head.

16. Diagrams generally should be given wherever they save time and help you to make a point clearly. Information presented in the diagram should not be repeated in words.

17. Diagrams should be large, clearly drawn in pencil and labelled fully, either in pencil or ink. Annotations should be added to the diagram if considered appropriate. Do not shade or colour diagrams unless absolutely essential to clarify what you have drawn. Label lines must touch the structure named.

18. In question papers that have printed leader lines or spaces for your answers, you should confine your answer to the space provided and not write in the margins or rule extra lines. The space given by the examiners is considered adequate for your answer.

19. Should you feel that you are running out of time, put down the important points in note form and do not worry about sentences. You may not gain full marks but you will gain some marks.

20. If time is available at the end of the examination read through your weakest answer and correct any mistakes.

1.5 Types of Questions

There are a variety of types of question papers and questions used in 'A' level examinations. Each type of question is designed to test specific skills and abilities by means of the candidate providing information on specific topics.

1. Essay Questions

This is a traditional type of examination question. It is open-ended and intended to give the candidate an opportunity to draw together information from various areas and present it in a logical way. No direction is given to the candidate as to how to structure the answer. It is therefore vital to draw up a plan before attempting to write the answer. This should be done on the examination paper alongside the question number. The plan should list the main points and key biological words to be contained in the answer, and the order in which they will appear. Answers must be concise, written in good English, and present relevant facts in a logical order. Remember to keep your sentences short. One of the skills tested by this type of question is the ability to communicate thoughts and ideas with clarity and precision, e.g.

> *Write an essay on the importance of ATP as an energy transfer molecule in the metabolic processes of living organisms.*   [O & C]

2. Structured, Free-response Questions

This type of question is intermediate between an essay question and a short-answer structured question. Diagrams should be included if they will help to clarify your answer. The question is usually in several parts. No guidance on length of answer is given and you should be guided by the mark allocation, e.g.

(a) Describe how
  (i) the action of the heart and
  (ii) the structure of blood vessels contribute to the unidirectional flow of blood and a regular pulse in a healthy person.   [8]
(b) How does the body detect and respond to a temporary fall in blood pressure?   [4]
(c) (i) How is tissue fluid formed and what is its composition and function?
  (ii) Explain how tissue fluid is eventually returned to the blood.   [8]
                                                                                        [JMB]

3. Structured, Short-answer Questions

Answers to questions of this type are to be written on the question paper in the spaces provided. This gives an indication of the length of answer required. *Conciseness* is important in this type of question. It is neither necessary nor desirable to provide answers outside the lines or spaces provided, e.g.

(a) State **six** main constituents of a fertile soil.

(i). Rock particles........... (iv)...Mineral salts.......

(ii)...Water............... (v)....Bacteria..............

(iii)...humus.............. (vi)..Earthworms....... [2]

Air sand clay

(b) Many soil fungi live as **saprophytes**. Define this term.

..A organism living on dead decaying matter by breaking down the complex organic matter [2]

(c) What is the importance of saprophytic fungi in the soil?

They release vital chemical elements into the soil., therefore aiding the recycling of material from dead to living [2]

(d) In what **two** other ways may fungi live in the soil?

(i) ..Parasitically as parasites.........

(ii) ..Symbiosis................ [2]

(e) State **two** important activities carried out by aerobic bacteria in the soil.

(i) Nitrogen fixation...............

(ii) Break down organic compounds to simpler ones. [2]

(f) In a fertile soil, as mineral ions are absorbed by plant roots they are continually being replenished. State **two** sources of supply.

(i) Decay of plant material........

(ii) Addition of fertilizers........ [2]

(g) *(i)* Clay soils often become waterlogged. Explain the effect this has on plants.

..Larger uptake of water., plant becomes very turgid. [2]

bad aeration, becomes cold + heavy 'bad growth'.

(ii) Clay soils often support luxuriant growth. Explain how this may happen.

clay can retain water and has a rich supply of minerals and if it has a good structure then luxuriant growth [3]

[Total 15]
[O & C]

4. Objective Questions

Answers to this type of question are usually made by placing a pencil line through a box or line on a special answer sheet. These sheets are marked by electronic scanner. All these questions have been pretested and are designed to test a variety of different levels of intelligence such as knowledge, comprehension, application, synthesis and evaluation. Between thirty and fifty questions are usually set—all are compulsory and they tend to cover most areas of the syllabus. It is advisable, therefore, to revise all the areas of the syllabus that you have covered in lessons. When answering an objective test-paper you should not spend too long working out a difficult question. If you cannot answer it quickly, leave it and go on to the next question. If you have time left at the end of the test, go back and work out the answer, e.g.

> The **secondary** order of protein structure is
> A the sequence of amino acids in the polypeptide chain.
> B the formation of peptide bonds between amino acids.
> C the coiling of the polypeptide chain.
> D the folding of the coiled polypeptide chain.
> E the linking together of two or more polypeptide chains.　　　　　　[UCLES]

Only one of the options (**A** to **E**) is correct. This is known as the key response. If you cannot tell which answer is correct you should try to eliminate the four which are incorrect. Of the four, one is usually almost correct, the major distractor. The other three are often very wrong. In the example above **B** is clearly wrong as this refers to the peptide linkage. **A** refers to the *primary* order. **E** describes a much higher order of structure, namely the *quaternary* order. This narrows the answer to **C** or **D**. Since **D** describes 'folding of a coiled peptide chain' it indicates a higher order than **C**. **C** is correct, as the secondary order of the protein structure describes the coiling of the polypeptide chain.

5. Practical Questions

These are designed to test your ability to follow instructions, to make accurate written or diagrammatic observations and produce valid conclusions. Time is an important factor in practical examinations. That does not mean that you need to rush. Rather, you should plan your work carefully and meticulously. You should read through the question paper carefully and plan how you will tackle each question. It is important that where apparatus needs to be set up and left for a while you should do this early in the examination. Remember that it is often necessary to take a number of readings in a physiological exercise so that time should be allocated for this.

Follow all instructions carefully. Your method should be clearly and logically presented so that your work could be repeated by the examiner. Volumes, weights, times and temperatures should be stated precisely. When asked to record your observations include all detail, e.g. size, shape, colour, smell etc.

Drawings should be large and clear and labelled as instructed. They should be accurate representations and be drawn according to the advice given on page 11. Plan diagrams of sections of plant and animal tissues should not include details of individual cells. Marks can only be awarded for information in your answer. Precision and conciseness are important.

Conclusions and deductions should be made on the basis of your observations and on your practical and theoretical knowledge acquired during the course.

1.6　　Terms Used in Examinations

During the preparation of an examination paper great care is taken with the wording of questions to ensure that they are concise and unambiguous. The majority of questions begin with an instruction designed to guide the candidate to provide the correct response. It is important that candidates should appreciate the meanings of these instructional terms.

1. **Annotated diagram**   Draw a large labelled diagram and alongside each label give a brief description of its structure and/or function as appropriate, e.g.:

   Make an annotated diagram to show the relationship (in terms of exchange of substances) between blood capillaries, tissue cells and lymph vessels.   [UCLES]

2. **Calculate**   Show all the stages involved in deriving the answer to a numerical problem. It is always a good idea to begin with a formula or an explanation of the terms used in the calculation, e.g.:

   Calculate the expected frequencies of the different genotypes in the population.   [UCLES]

3. **Compare**   State the similarities and differences between the two or more topics stated in the question, e.g.:

   Compare gaseous exchange in the leaf of a flowering plant with that in the mammalian lung.

   If comparisons are being made between, say, A and B, you should refer to both, e.g. 'A has . . . . . . whereas B has . . . .' In the case A would be the leaf and B would be the lung.   [O & C]

4. **Contrast**   State the differences between two or more topics, e.g.:

   Contrast the actions of the nervous system and endocrine system in mammals.

5. **Compare and contrast**   State, point by point, the similarities and differences between two or more subjects, e.g.:

   *Compare and contrast the process of osmoregulation in a **named** marine fish and a **named** fresh-water fish.*

   It is vital to plan your answer to this type of question. One way of doing this is to list the main features of the process and the similarities and differences shown by each of the named examples; i.e., for the question above:

	Marine fish	*Fresh-water fish*
Environment	Hypertonic	Hypotonic
Point of influx/outflux		
etc.	Gills	Gills

6. **Define**   State briefly the meaning of the term, e.g.:

   Define an enzyme in chemical terms.   [UCLES]

7. **Describe**   State in words (using diagrams where appropriate) the main points of the structure or process in the question. This is a very common instruction on examination papers and it requires you to impart knowledge of facts or processes to the examiner. It is a request to you for information. As usual, answers should be concise and presented logically, e.g.:

   Describe how plants and animals get rid of their waste products.   [L]

8. **Discuss**   Give a *critical* account of *all* of the points involved in the topic being written about, and their relative significance and importance. It is wise to list these points in your plan and present them in an orderly way in your answer, e.g.:

   Discuss man as an ecological factor.   [O & C]

9. **Distinguish between**   State the *differences* between the subjects mentioned in the question and their implications. It is useful to state a named example to illustrate your answer, e.g.:

   *Distinguish between a taxis and a tropism, giving **two** examples of each.*   [O & C]

10. **Diagrams** Specific questions may be set on diagrams or they may be used to illustrate a written answer. In all cases diagrams should be large, fully labelled and have a title. Simple pencil line drawings are all that is needed and they may be labelled in pencil or ink. Shading and colouring should be avoided as they usually do not aid clarity. Labelling lines should touch the appropriate structure and should be arranged neatly around the diagram. Lines must not cross. Correct proportion is important, e.g.:

Draw a labelled diagram of a generalized animal cell to indicate the distribution of the membranes. [UCLES]

11. **Explain** State, with reference to theory and reason and cause and effect, the implications of the factors affecting the subject. Clarity and relevance are important in answers to this type of question, e.g.:

Explain the ways in which green plants are dependent on the activities of animals. [L]

12. **Graph** It is important that you should be able to plot graphs accurately and interpret data presented in the form of a graph.
   (a) Graphs should always be drawn in pencil on graph paper.
   (b) The graph should fit the middle of the graph paper.
   (c) The horizontal x-axis should represent the magnitude of the independent variable, that is the variable whose value is chosen by the experimenter, e.g., time, temperature.
   (d) The vertical y-axis should represent the magnitude of the dependent variable, that is the variable whose value is not chosen by the experimenter.
   (e) Each axis should begin at 0. If the values on one axis are grouped between, say, 90 and 100, a large scale will be required. Still begin the axis at 0 but make a break in the axis, marked as $-/\,\vdash$, at a point just beyond 0.
   (f) Axes must be fully labelled in terms of the variable, e.g. temperature/°C, and its values should be equally spaced.
   (g) Points on the graph should be clearly marked by an x or a circled dot, never by a dot alone.
   (h) Points may be joined either by a series of straight line segments drawn with a ruler, or with a smooth curve.
   (i) The lines on a graph should be referred to as either lines or curves.
   (j) If several curves are drawn on the same axes, each must be labelled clearly.
   (k) The graph should have a title, such as '*Graph showing the relationship between . . . .*'.
   (l) If asked to interpret the graph, you should relate the changes shown by the line(s) or curve(s) to the biological situation illustrated. First describe what happens to y-values as a result of changes in x-values. The next step is to apply these trends to the biological context. Remember too, that the steeper the slope, the faster the change.

13. **Illustrated account** Diagrams should be used as much as possible in the answer. These may be annotated, flow or structural diagrams. Writing should only be used to explain points which cannot be made in other ways, e.g.

Give an illustrated account of the way in which skeletal and muscle systems operate together in a named terrestrial mammal to produce rapid locomotion. [O & C]

14. **List** Implies brevity. The facts should be numbered 1, 2, 3, etc. and stated as briefly as possible, e.g.:

List the problems faced by an animal living on dry land. [UCLES]

15. **Outline**   This also implies brevity. Only absolutely essential information should be stated, e.g.:

Outline the chemical changes which proteins and fats undergo when they are used as an energy source. Indicate where the products of these changes link up with carbohydrate respiration.   [NISEC]

16. **State**   This too requires a *concise* answer. Information should be stated without recourse to supporting reason, e.g.:

State the essential difference between prokaryotic and eukaryotic cells, and give an example of each type of cell.   [UCLES]

17. **Suggest**   This is usually used to imply that you should state your answer on the basis of theoretical knowledge. It is not anticipated that you will 'know' the answer. There may not even be a correct answer, e.g.:

*Suggest **two** ways in which the reproductive tract of a female mammal can be said to be a hostile environment for sperm.*   [AEB]

18. **What is meant by** the term(s) . . . .   Normally implies that a definition is required, supported by relevant comment on the significance or context of the term(s) used. The amount of additional information given in your answer should be dictated by the mark value indicated, e.g.

What is meant by 'alternation of generations'?   [L]

2 Chemicals of Life

Condensation   The reaction occurring when two molecules are joined together to form a larger molecule with the subsequent loss of a water molecule.

Hydrolysis   The reaction occurring when one molecule is broken down into two molecules by the addition of water.

Macromolecule   A large molecule built up from many repeating smaller units.

Isomerism   The situation which exists where two different compounds possess the same molecular formula.

Ester   An organic compound formed by a reaction between an acid and an alcohol.

Iso-electric point   The pH at which amino acids and proteins exist in their neutral (zwitterion) form.

Amphoteric   A compound possessing both basic and acidic properties.

Protein conformation   The characteristic three-dimensional shape of a protein.

Enzyme   A protein catalyst produced by a living organism.

Activation energy   The least amount of energy required for a reaction to occur.

Cofactor   A small non-protein molecule required for the activity of some enzymes.

2.1   Basic Biochemistry

The four most common elements in living organisms are hydrogen, carbon, oxygen and nitrogen.

Carbon has the following important properties:

1. It forms four stable covalent bonds.
2. It can form carbon–carbon bonds which can be used to build carbon skeletons with chain and/or ring structures.
3. It can form multiple covalent bonds with carbon, hydrogen or nitrogen atoms.

Water is the most abundant simple compound in living organisms. It possesses the following biologically important characteristics and features:

1. It is an excellent **solvent**, because of its polar properties. It transports solutes within cells and between cells.
2. Its **high specific heat** enables living matter to undergo extensive heat absorption or loss with only a minimal temperature change. Therefore enzyme reactions can proceed within a narrow temperature range and at a relatively constant rate.
3. Its **high heat of vaporization** means that large quantities of heat energy are needed to vaporize a small quantity of water. This provides the basis of a very efficient cooling process, e.g. transpiration in plants and sweating in animals.

4. Water is **less dense** as a solid than as a liquid. Therefore ice floats, and life can continue in the liquid water below. Considerable heat has to be lost before ice forms, and cell contents are consequently less likely to freeze.

5. Water molecules have a **high surface-tension**. Hence they can associate with each other (cohesion) and may stick to (adhesion) a variety of different surfaces. This is of considerable importance when water is transported through the xylem in plants.

6. Water takes part directly or indirectly in all **metabolic reactions** occurring in living matter, e.g. condensation, hydrolysis, photosynthesis and respiration.
   Water has a pH of 7. It is a **neutral** liquid medium.

2.2   Organic Molecules

Simple organic molecules present in living material include monosaccharides, amino acids, fatty acids, glycerol and aromatic bases. Called **monomers**, they act as building blocks for the synthesis of larger **macromolecules**, called **polymers**, by **condensation** reactions.

$$\text{monomer} \rightleftharpoons \text{polymer}$$

e.g.   monosaccharides $\rightleftharpoons$ polysaccharides

(a) Carbohydrates

Carbohydrates are composed of the elements carbon, hydrogen and oxygen. The general formula is $C_x(H_2O)_y$.

Monosaccharides, e.g. glucose, fructose, galactose, are the simple 'sugars'. Each has the chemical formula $C_6H_{12}O_6$, but the arrangement of atoms and chemical groups in them is different. Because of this they exhibit **isomerism**. Monosaccharides are classified according to the number of carbon atoms they possess:

3C trioses   e.g. glyceraldehyde, dihydroxyacetone (involved in respiration);

5C pentoses   e.g. ribose (involved in the synthesis of nucleic acid, coenzymes and ATP);

6C hexoses   e.g. glucose, fructose (major sources of energy; involved in synthesis of polysaccharides).

Pentose and **hexoses** are the most common monosaccharides and exist as chain or ring structures. Ring structures are the usual forms found in nature. The 5-membered pentose ring is called a **furanose** ring and the 6-membered hexose ring is called a **pyranose** ring.

When two monosaccharides are joined together they form a **disaccharide** (see Fig. 2.1), e.g. maltose.

The following common disaccharides are formed from the monomers:

glucose + glucose   = maltose
glucose + galactose = lactose
glucose + fructose   = sucrose

Polysaccharides (e.g. starch, glycogen, cellulose) are formed when large numbers of monosaccharides are joined together in long chains. They provide structural support in plants or act as food and energy stores, e.g. starch in plants and glycogen in animals. They are relatively insoluble in water and have no osmotic effect within the cell.

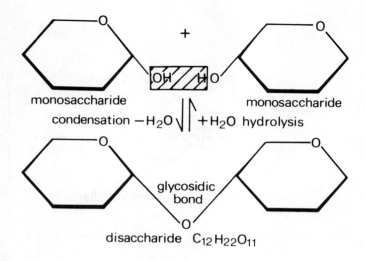

monosaccharide   monosaccharide

condensation $-H_2O$ $+H_2O$ hydrolysis

glycosidic bond

disaccharide $C_{12}H_{22}O_{11}$

Fig. 2.1

Starch is composed of amylose, a straight chain of several thousand glucose residues, and amylopectin, a larger branched structure containing many more glucose residues. Starch accumulates as starch grains in many plant cells and is frequently their major carbohydrate store, e.g. potato tubers.

Glycogen is the animal store of carbohydrate. It is a branched structure of thousands of glucose residues. In vertebrates it is characteristically found as tiny granules in liver and muscle cells.

Cellulose is the structural component of plant cell walls. It consists of long chains of glucose residues. Hydroxyl groups project outwards to form hydrogen bonds with hydroxyl groups of adjacent chains. This gives the whole structure a high tensile-strength. Cellulose is fully permeable to water and solutes and therefore does not affect exchange of materials between the cell and environment. Symbiotic bacteria in the gut of ruminant mammals digest cellulose into glucose molecules thus making it available for absorption.

Chitin is structurally very similar to cellulose, and forms bundles of long parallel chains which are an essential part of the arthropod skeleton.

(b) Lipids

Lipids are esters of fatty acids and glycerol. Fatty acids have the general formula RCOOH. R is an H or alkyl group. Most fatty acids have an even number of carbon atoms, usually between 14 and 22. These form a long hydrocarbon tail which is hydrophobic. Unsaturated fatty acids contain one or more double bonds (C=C) while saturated fatty acids lack double bonds. Triglycerides are the most common lipids. They are formed when each of the three hydroxyl groups of glycerol condenses with a fatty acid (Fig. 2.2). The fatty acids may be the same or different. A fatty acid may be substituted by a phosphate group, when a phospholipid is formed. Phospholipids are important structural components of plasma membranes.

(i) *Triglycerides*

These are called **fats** if they are solid, or **oils** if they are liquid, at a temperature of 20 °C. They store twice the amount of energy as the equivalent mass of carbohydrate and act as important energy reserves in animals and plants. Extra fat is

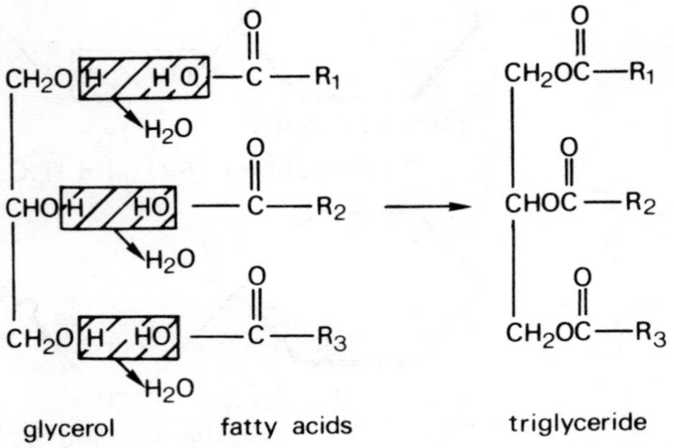

Fig. 2.2

stored by hibernating animals and below the dermis of the skin in mammals where it helps insulate the body (e.g. blubber in whales). It also surrounds and protects various vital organs in the mammalian body e.g. heart and kidney. When fats are respired water is one of the by-products. Called metabolic water, it is the main source of water for a number of desert animals, e.g. the kangaroo rat.

(ii) *Steroids*

These include the sex hormones, the adrenal cortex hormones and cholesterol. Cholesterol is present throughout the human body. It is an important component of cell membranes and a constituent of bile which emulsifies fat. It also acts as an intermediate for the synthesis of other steroids.

(iii) *Waxes*

These act as protective waterproof covering on bird feathers, mammal fur and skin, arthropod exoskeletons and the leaves and fruits of plants.

(c) Proteins

Proteins contain the elements carbon, hydrogen, oxygen, nitrogen and sometimes sulphur. They are made up of long chains of α-amino acids. The general formula of an **amino acid** is:

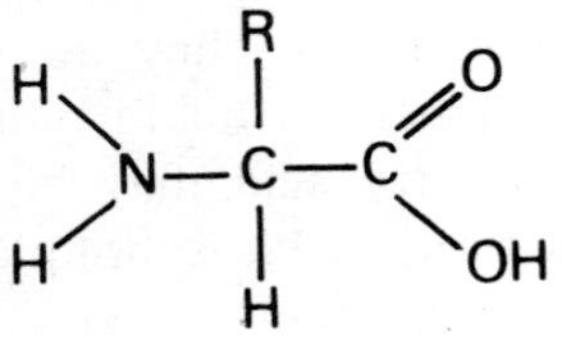

The simplest amino acid is glycine. It is formed when the R group is substituted by H. Amino acids are soluble in water and amphoteric. This is a useful biological property as it means they can act as buffers in solutions resisting wide fluctuations in pH (Fig. 2.3).

A **dipeptide** is formed when two amino acids are joined together by a condensation reaction (Fig. 2.4).

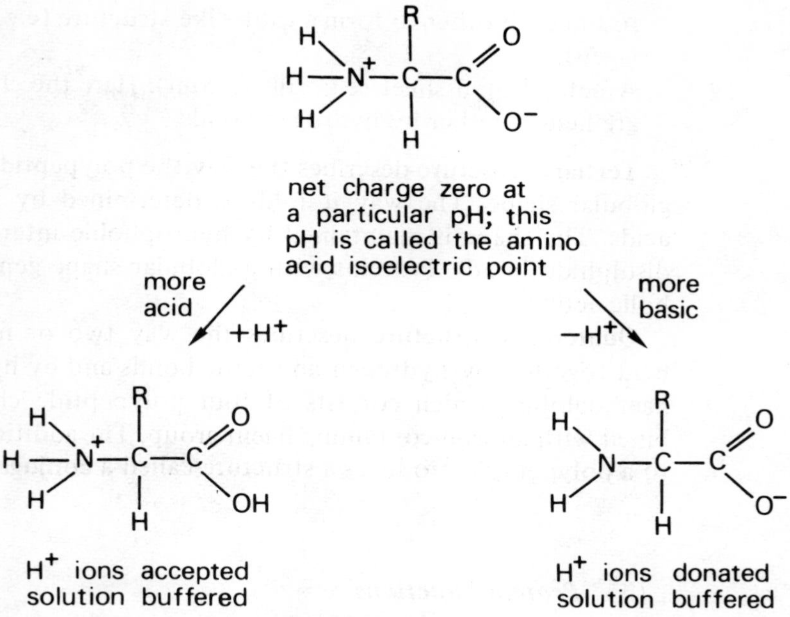

net charge zero at
a particular pH; this
pH is called the amino
acid isoelectric point

more
acid $+H^+$

more
basic $-H^+$

H$^+$ ions accepted
solution buffered

H$^+$ ions donated
solution buffered

Fig. 2.3

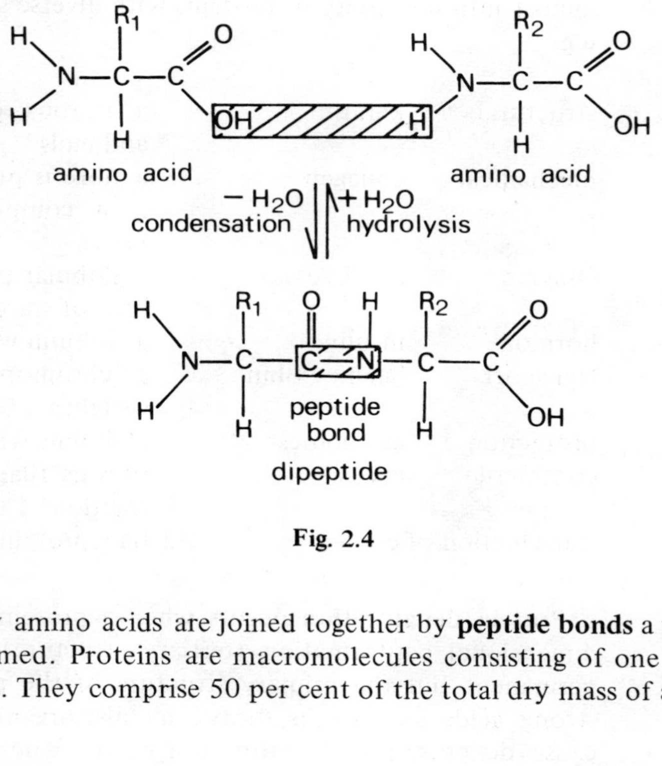

amino acid   amino acid

$-H_2O$ $+H_2O$
condensation   hydrolysis

peptide
bond

dipeptide

Fig. 2.4

When many amino acids are joined together by **peptide bonds** a **polypeptide** or protein is formed. Proteins are macromolecules consisting of one or more polypeptide chains. They comprise 50 per cent of the total dry mass of animal cells.

(i) *Protein Structure*

The **primary structure** of proteins is the linear sequence of amino acids in a polypeptide chain. The sequence largely determines the protein's biological function. For this reason proteins are regarded as informational macromolecules.

Secondary structure is the arrangement of the primary structure along one axis. The fibrous structures may form:

1. A helix which is held in place by hydrogen bonds between adjacent CO and NH groups (e.g. the fibrous protein keratin). Sometimes several chains wind

round each other to form a cable-like structure (e.g. the triple helix of tropocollagen).

2. A beta-pleated sheet (e.g. silk fibroin). Here the chains line up side by side and are held together by hydrogen bonds.

Tertiary structure describes the way the polypeptide chain folds into a compact globular shape. The way it folds is determined by interactions between amino acids. The shape is maintained by hydrophobic interactions, ionic, hydrogen and disulphide bonds. Proteins with a globular shape generally possess specific metabolic activity.

Quaternary structure describes the way two or more polypeptide chains are held together by hydrogen and ionic bonds and by hydrophobic interactions, e.g. haemoglobin, which consists of four polypeptide chains each of which is combined with an iron-containing haem group. The addition of a non-protein molecule to a polypeptide produces a structure called a **conjugated protein**.

(ii) *Protein Functions*

The functions of proteins depend upon the nature and arrangement of the constituent amino acids. Twenty different amino acids commonly occur in proteins. They are organized into many different sequences which in turn give rise to an almost infinite variety of proteins with diverse structural and metabolic functions, e.g.

structural	keratin	a fibrous protein found in skin, feathers and nails
mechanical	collagen	a fibrous protein which resists stretching and is a component of tendons, cartilage and bones
enzyme	amylase	a globular protein which catalyses the hydrolysis of starch to maltose
hormone	insulin	a globulin which aids glucose metabolism
transport	haemoglobin	a chromoprotein, it transports oxygen in vertebrate blood
protection	antibodies	globulins which neutralize foreign proteins
contractile	myosin	fibrous filaments involved in sarcomere contraction
transduction of energy		flavoproteins—FAD

Protein denaturation occurs when a protein molecule loses its specific three-dimensional conformation (tertiary structure). The loss may be temporary or permanent but the primary structure of the protein remains unaffected. Heat, strong acids and alkalis, heavy metals, organic solvents and detergents can all cause denaturation. Renaturation occurs when a protein spontaneously refolds into its characteristic 3 D form after denaturation.

(d) Vitamins

Vitamins are complex organic compounds synthesized by plants and animals. Names of vitamins, sources, functions and symptoms of deficiency are not described here. You should refer to a standard text for this information.

(e) Nucleic Acids

Nucleic acids are informational macromolecules and constitute the genetic material of all living organisms. They are made up of subunits called **nucleotides**. Each nucleotide possesses three components, a 5-carbon sugar, a nitrogenous base and phosphoric acid.

(i) DNA

This consists of two helical anti-parallel polynucleotide chains held together by the pairing of complementary bases between adjacent chains (Fig. 2.5). Adenine is always paired with thymine by two hydrogen bonds, and guanine is always paired

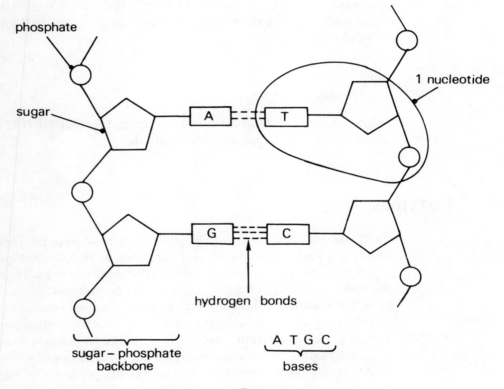

Fig. 2.5

with cytosine by three hydrogen bonds. These four bases constitute the genetic alphabet used to spell out genetic instructions. Each polynucleotide chain has a sugar-phosphate backbone with bases projecting at right angles from it. The sequence of bases in one chain is entirely random, but because base pairing occurs, the sequence on one chain determines that on the other. In other words the chains are complementary to one another. This makes them ideally suited for accurate copying of the genetic material.

(ii) RNA

This contains the sugar ribose in its structure while DNA contains deoxyribose. Each nucleotide possesses one of four different bases. The bases are the purines adenine (**A**) and guanine (**G**), the pyrimidines thymine (**T**) in DNA or uracil (**U**) in RNA, and cytosine (**C**).

19

RNA is normally single-stranded. Uracil replaces thymine. RNA exists in three forms:

(i) *Messenger RNA*

This is a long molecule which carries the DNA message from the nucleus to the cytoplasm in the form of a series of codons (sequences of three nucleotides). This process is called transcription.

(ii) *Transfer RNA*

A small molecule, cloverleaf in shape, links with a specific amino acid at one end, while an anticoden at its other end fits on to the mRNA at a complementary mRNA codon.

(iii) *Ribosomal RNA*

This is located in the ribosomes and brings together mRNA, tRNA and associated amino acids during polypeptide synthesis.

2.3 Enzymes

Enzymes are complex protein molecules produced by living organisms. They catalyse a vast range of chemical reactions. If this did not occur reactions in the cell would be too slow to support life. Generally, a sequence of different enzymes are linked together. This is called a metabolic pathway.

A linear pathway has a number of enzymes, usually membrane-bound, arranged sequentially as a multi-enzyme complex. Collectively they produce a specific end-product. Self-regulation may occur by the end-product inhibiting a reaction occurring earlier in the complex. This is a form of negative feedback.

Alternatively, a branched pathway has the potential to form a number of end-products according to the conditions prevailing in the cell at a given time. These enzymes are not usually associated with one another and are generally found in solution.

(a) Properties of Enzymes

1. All are globular proteins.
2. They increase the rate of a reaction without themselves being used up.
3. Their presence does not alter the nature of the end-products of the reaction.
4. A small amount of enzyme catalyses a large quantity of substrate.
5. Their activity varies with changing pH, temperature, and substrate concentration.
6. They generally only catalyse one specific reaction.
7. They catalyse a reaction in either direction according to prevailing conditions.
8. They reduce the activation energy (E_a) required for a chemical reaction to take place.

(b) Mechanism of Enzyme Action

In the **lock-and-key theory** the substrate fits exactly into a small portion of the enzyme called the **active site**. The substrate is often called the 'key' whose shape complements exactly the enzyme or 'lock'.

In the **induced-fit theory** the bonds between the amino acids of the active site of the enzyme are relatively flexible. When a substrate combines with an enzyme, the active site may mould into the shape of the substrate.

Cofactors are non-protein components required by enzymes for their efficient functioning. An enzyme–cofactor complex is called a holoenzyme. An enzyme without its cofactor is called an apoenzyme. There are three types of cofactor:

1. **Inorganic ions**; e.g. salivary amylase activity is increased by the presence of chloride ions.
2. **Prosthetic groups**. These are non-protein organic molecules tightly bound to the active site, e.g. FAD, haem.
3. **Coenzymes**: non-protein organic groups loosely associated with the enzyme, e.g. NAD, ATP.

Prosthetic groups and coenzymes act as carriers of groups of atoms.

(c) Enzyme Inhibition

A number of small molecules can reduce the rate of an enzyme-controlled reaction:

1. **Competitive reversible inhibition**: a compound similar in structure to the usual enzyme substrate competes for a place on the enzyme's active site. The reaction can be reversed if the substrate concentration is increased.
2. **Irreversible inhibition**: small concentrations of heavy-metal ions combine permanently with the enzyme and completely inhibit it.

(d) Allosteric Enzymes

Compounds called allosteric effectors bind to these enzymes at sites away from the active site for the substrate. They may speed up or slow down the enzyme catalysed reaction.

(e) Classification of Enzymes

Enzymes are placed in groups according to the general type of reaction they catalyse: see Table 2.1.

Table 2.1

Type of reaction		Example
oxidoreductase	(i) $AH + B = A + BH$	dehydrogenase
	(ii) $A + O = AO$	oxidase
transferase	$AB + C = A + BC$	transaminase
hydrolase	$AB + H_2O = AOH + BH$	amylase
lyase	$\underset{\underset{OH}{\overset{\|}{}}}{R-C-\overset{\overset{O}{\|}}{C}} \rightleftharpoons \underset{\underset{H}{\overset{\|}{}}}{R-\overset{\overset{O}{\|}}{C}} + CO_2$	decarboxylase

2.4 Worked Examples

2.1

(a) Give the general formula for a carbohydrate.

$C_x(H_2O)_y$ .... [1]

The carbohydrate illustrated below has been formed from two hexose sugars:

(b) What type of carbohydrate is this? .... *disaccharide* .... [1]

(c) Name the type of chemical bond which joins the two hexose units together in the molecule. ... *1, 4 glycosidic bond*

.... [1]

(d) What is the name given to the chemical reaction in which two or more hexose sugars combine to form larger units?

condensation reaction .... [1]

(e) Give one function of the carbohydrate illustrated above in living organisms.

acts as an energy-rich substrate .... [1]

(f) Name a storage polysaccharide commonly found in mammals.

glycogen .... [1]

(i) In which organ of the body would you expect to find it occurring in relatively large amounts? *liver* .... [1]

(ii) Where is it deposited in the cells of this organ?

in glycogen (zymogen) granules .... [1]

(iii) In what physical form is this substance found in cells?

insoluble .... [1]

[OLE]

2.2

(a) Give an account of the chemical nature and variety of carbohydrates. **[10]**

(b) Outline the role of carbohydrates in the life of a plant. **[8]**

[UCLES]

(a) Carbohydrates are composed of the elements carbon, hydrogen and oxygen. The elements are joined together by covalent bonds to form either aldehydes or ketones. The groups determine the chemical nature of the compounds.

22

There are three major groups of carbohydrates:

(i) Monosaccharides: these are made up of single 'sugar' units and classified according to the number of carbon atoms, e.g. trioses (3C), pentoses (5C) and hexoses (6C). Each monosaccharide exists either as an aldose or as a ketose sugar, e.g. glucose and fructose respectively.

(ii) Disaccharides: these are formed by a condensation reaction between two monosaccharide units. The bond between them is a glycosidic bond.

(iii) Polysaccharides: these are polymers of mono- and disaccharide units. They may be branched as in cellulose or unbranched as in amylose.

Both monosaccharides and disaccharides are crystalline, readily soluble in water and sweet. They can exist as α or β isomers and this gives rise to greater chemical variety. All monosaccharides and some disaccharides are reducing sugars. Sucrose is a non-reducing sugar. Polysaccharides have high relative molecular masses and are non-crystalline, insoluble and not sweet. The structure of polymers is variable and gives the molecules compactness, e.g. starch, and strength, e.g. cellulose.

(b) Glucose is a universal respiratory substrate. It is the fundamental molecule for the formation of all other carbohydrates, as well as for proteins as fats, nucleic acids and co-enzymes.

Maltose and sucrose are important energy sources and can be hydrolysed to monosaccharides. Sucrose is an important sugar in plants where it is translocated through the phloem. It is an important component of vacuolar cell sap where it exerts a low water potential. The resulting inward flow of water produces an outward pressure which gives plant cells turgidity.

Starch is a polymer of glucose. In plants it forms helices and acts as a storage compound, which on hydrolysis yields monosaccharides for respiration. Cellulose consists of long chains of glucose residues which form hydrogen bonds with neighbouring chains. These rigid cross-links form structures with high tensile strengths. This is important for the structural role played by cellulose in plant cell walls. Cellulose is an important food source of bacteria fungi and many animals possessing the enzyme cellulase.

2.3

(a) Give an account of the structure of proteins. [10]

(b) How are the functions of proteins related to their structure? [8]

[UCLES]

(a) The information required to answer this question is provided in section 2.2. The format of the answer presented should be as for Worked Example 2.2.

(b) The information required to answer this question is provided in section 2.2. The answer should include the various functions of proteins along with examples. The ways in which these structures influence the functions stated should be described.

2.4

(a) Name one food material rich in vitamin A.

milk

.... [1]

(b) Where in the mammalian body is vitamin A stored? ....

liver

.... [1]

(c) What is the function of vitamin A? . . . .

controls normal epithelial structure and growth
. . . . **[1]**

(d) State one function of vitamin B_{12} in man. . . . .

erythrocyte formation
. . . . **[1]**

(e) Which trace element is present in vitamin B_{12}? . . . .

cobalt
. . . . **[1]**

(f) Name a disease associated with defective absorption of vitamin B_{12}.

pernicious anaemia
. . . . **[1]**

[OLE]

2.5

(a) In the DNA molecule there are four major bases, two purines and two pyrimidines. These are

purine   *(i)*   adenine (A) . . . .

                  (ii)   guanine (G) . . . .

pyrimidine   *(i)*   thymine (T) . . . .

                  (ii)   cytosine (C) . . . .

                                                               [4]

(b) Name the sugar in

    (i) DNA   deoxyribose

    (ii) RNA   ribose

                                                                [2]

(c) What is a nucleotide?

a compound composed of a sugar, a nitrogenous
base and a phosphate
                                                                [2]

(d) What are the possible base pairings of the four bases?

A with T
C with G
                                                                [2]

(e) One of the bases in DNA is not represented in RNA. Name the base not represented and the base that replaces it.

    (i) Base not represented   thymine

    (ii) Base that replaces it   uracil
                                                                [2]

(f) Name three types of RNA involved in protein synthesis.

    (i)   messenger RNA (mRNA)

    (ii)   transfer RNA (tRNA)

    (iii)   ribosomal RNA (rRNA)
                                                                [3]

[L]

2.6

The diagram shows part of a polynucleotide chain.

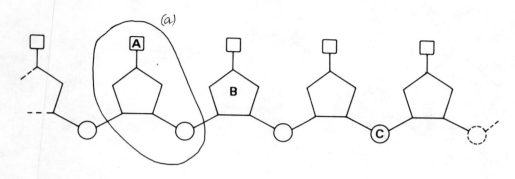

(a)

(a) On the diagram draw a line round **one** nucleotide.

(b) What are the chemical groups labelled A, B, C?

A .. base ....

B .. deoxyribose sugar ....

C .. phosphate ....

(c) Which compound pairs with thymine in the DNA molecule?

adenine ....

(d) How are the bases on the two adjacent strands of the DNA molecule held together?

hydrogen bonding

....

[6]
[AEB]

2.7

For each of the following formulae give (i) the name of the substance represented and (ii) the biochemical class to which it belongs.

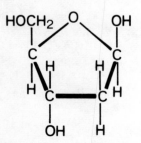

Name of substance	Biochemical class
uracil	pyrimidine

....

deoxyribose   pentose

....

continued

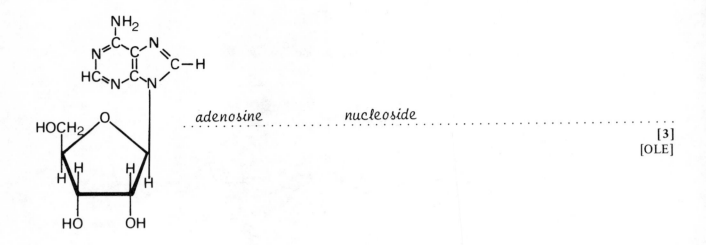

adenosine   nucleoside ....
.... [3]
[OLE]

2.8

(a) *Briefly explain what you understand by the term enzyme.*
A protein catalyst produced in organisms which increases the
rate of chemical reaction without being used up in the reaction.
.... [2]

(b) *State two factors other than temperature which influence the rate of enzyme activity.*
(i) pH ....
(ii) substrate concentration ....
.... [2]

(c) *Explain briefly the meaning of the following terms in relation to enzyme structure
and activity:*
(i) prosthetic group; a non-protein organic molecule tightly bound to
the enzyme at all times
.... [2]
(ii) co-enzyme; a non-protein organic molecule loosely bound to an
enzyme. It binds substrate and enzyme.
.... [2]
(iii) non-competitive inhibition; binding of inhibitor molecule to enzyme
so as to alter shape of active site and reduce activity
.... [2]
(iv) active site. the region of the enzyme molecule which combines
with the substrate
.... [2]
[OLE]

2.9

(a) *Give an example of one plant and one animal enzyme that act intracellularly and state
their functions.*
plant enzyme ribulose bisphosphate carboxylase ....

26

function .... *catalyses the reaction between ribulose bisphosphate,* ....

carbon dioxide and water to form PGA

animal enzyme .... *acetylcholinesterase* ....

function .... *breakdown of acetylcholine to choline and*

acetic acid

[4]

(b) *Give an example of one plant and one animal enzyme that act extracellularly and state their functions.*

plant enzyme .... *cellulase*

function .... *promotes breakdown of cellulose to glucose*

animal enzyme .... *maltase*

function .... *catalyses the hydrolysis of maltose to glucose*

[4]
[L]

2.10

The following passage refers to enzymic reactions. Read the passage carefully and then answer the questions that follow.

'Three major events occur during an enzymic reaction. In the first critical step free enzyme and free substrate bind to each other forming the enzyme-substrate complex. The complex is then activated. The role of enzymes is to accelerate chemical transformations by reducing the energy barriers to chemical reactions. These barriers are the primary determinants of how rapidly the chemical reactions of metabolism occur at biological temperatures. The rate at which a chemical reaction occurs is largely set by the free energy of activation of the reaction. The catalytic function of enzymes is to reduce the amount of this input of energy which is required to form the active complex.*

'Once the active complex is generated, the third event, the conversion of substrate(s) to product(s) is usually very rapid. Following product formation, the enzyme must free its substrate binding site. Thus, just as enzymes must efficiently bind substrate in the initial step of the reaction, they must be able to release the product(s) of the reaction. Biochemists often speak of "turnover" of substrate molecules and express the speed of the reaction as its "turnover number". The higher the turnover number of an enzyme, the greater its efficiency.'

(a) *What two features of an enzyme and its substrate are of critical importance if the enzyme substrate complex is to be formed?*

(i) *relative concentrations of enzyme and substrate*

(ii) *The shape of the substrate molecule should be compatible with the enzyme's active site.*

(b) *Explain what is meant by*

(i) *biological temperatures (line 5);*

(ii) *free energy of activation (line 6).*

(i) *the temperature range over which enzymes function most efficiently, e.g. 20-40°C.*

(ii) ......the energy required by the enzyme/substrate complex in
order for the substrate to break down to form products ......

(c) *Why must an enzyme free its substrate binding site? (lines 10–11)*
   to allow further reactions to occur
......

(d) *How would you define 'turnover number'?*
   This is the number of substrate molecules broken
   down per unit time per unit amount of enzyme. ......

.... [7]

2.11

Draw simple graphs which illustrate how the rate of a typical enzyme-controlled reaction varies with (a) temperature; (b) pH; (c) substrate concentration; (d) concentration of a competitive inhibitor. In each case, explain as fully as you can the reasons for the relationship.
[5, 5, 4, 4]
[UCLES]

(a)

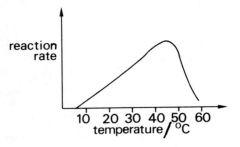

The majority of mammalian enzymes have an optimum temperature of about 35 to 40 °C. At higher temperatures the enzymes denature and all activity ceases at about 60 °C, the temperature at which most proteins coagulate. Temperature influences the rate of a reaction by increasing molecular motion. The reactants are therefore more likely to meet as well as attach to the active site of the molecule. In enzyme reactions the temperature coefficient, Q_{10}, of the reaction is about 2 over the range of 0 to 40 °C.

$$Q_{10} = \frac{\text{rate of reaction at } (x + 10\,°C)}{\text{rate of reaction at } x}$$

At temperatures in excess of 40 °C the secondary and tertiary enzyme structures denature. The specific configuration of the active site is destroyed and enzyme activity ceases. At low temperatures, i.e. less than 10 °C, enzymes are inactivated but not denatured.

(b)

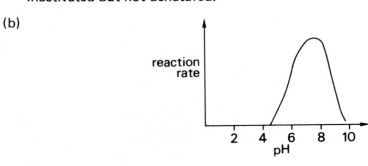

At a constant optimum temperature all enzymes function most efficiently over a narrow pH range. pH affects the ionizable groups of the enzyme, especially those at the active site. This causes a change in the molecular configuration of the active site and affects its ability to form a substrate-enzyme complex. The net effect is a reduction in catalytic activity. Extreme changes in pH alter the ionic charges on acidic and basic groups of peptide chains in the enzyme molecule. They can also destroy the secondary and tertiary protein structures by breaking hydrogen bonds and disulphide bridges.

(c)

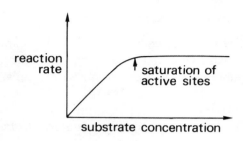

For a given enzyme concentration the rate of an enzyme-mediated reaction increases with increasing substrate concentration in a linear fashion. Above a certain substrate concentration the rate of reaction becomes constant. This is because at this point all the active sites, at any moment, are fully saturated with substrate. Further substrate reactivity must await dissociation of the enzyme–substrate complex. Substrate concentration can therefore limit reaction rate. The rate can be increased above this level by increasing the enzyme concentration.

(d) A competitive inhibitor has a molecular shape similar to that of the substrate. The competitive inhibitor will compete with the substrate for attachment to the active sites. This reduces the number of enzyme–substrate complexes that can form. For fixed amounts of substrate and enzyme the graph of reaction rate against concentration of inhibitor will appear as here:

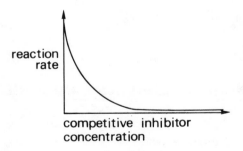

The curve shows that at low concentrations of competitive inhibitor the reaction rate is high, providing substrate concentration is in excess. As the concentration of inhibitor increases, the rate of enzyme activity decreases due to the competitive inhibitor blocking the active sites. Even at high concentrations of inhibitor there will still be some enzyme activity.

3 Cytology and Histology

> **Cytology**   The study of cells by means of light and electron microscopy.
> **Cell Theory**   All living matter is composed of cells; all cells arise from pre-existing ones; all metabolic reactions occur within cells.
> **Protoplasm**   The living contents of the cell consisting of the cytoplasm (including the cell membrane) and one or more nuclei.
> **Organelle**   A structurally and functionally discrete component of a living cell, frequently bounded by a membrane.
> **Histology**   The study of the structure of tissues by means of light and electron microscopy.
> **Tissue**   A collection of structurally similar cells specialized to perform a particular function or functions.
> **Organ**   A distinct structural and functional unit of the body composed of more than one tissue.

3.1   Cytosol

This is the fluid part of cytoplasm. It contains the biochemicals necessary for life and is the site of some important metabolic pathways, e.g. glycolysis in respiration. Cytoplasmic streaming occurs which aids movement of materials throughout the cell.

3.2   Cytoplasmic Organelles Common to Animal and Plant Cells

(a) Cell Membrane

The **'fluid mosaic' model** of membrane structure states that there is a central dynamic bilayer of lipids containing freely floating proteins, sandwiched between two protein layers. The proteins are variable in function and can act as enzymes, receptors, electron carriers and energy transducers. The membrane (known as the unit membrane) is of the same general structure as the membranes of other organelles; see the answer to Worked Example 3.7. It is **differentially permeable** and controls the exchange of materials between the cell and its environment. In this way it maintains an optimum pH and ionic balance for enzyme activity, excretes waste, secretes useful materials and generates ionic gradients for muscular and nervous activities. Materials move through the membrane by diffusion, osmosis, endo/exocytosis or by active transport which involves ionic pumps driven by ATP.

(b) Nucleus

The nucleus is separated from the cytoplasm by a **nuclear envelope** that is perforated by **nuclear pores** through which materials are exchanged. Chromosomes, composed of DNA and protein, are present in the **nucleoplasm**. DNA controls the cell's activities, can replicate, and is involved in cell division. In a non-dividing cell chromosomes exist in an elongated form called **chromatin**. Several portions of different **chromosomes**, called nuclear organisers, meet at the well-defined **nucleolus** within the nucleus. They contain the **genes** needed by the nucleolus for the manufacture of ribosomal RNA.

(c) Endoplasmic Reticulum (ER)

This is a network of membranous sacs called **cisternae**, arranged in tubes or sheets. It is the site for many metabolic activities. It also channels the movement of materials and separates different activities of the cell which are proceeding simultaneously. **Rough ER** has ribosomes attached to its surface and is the site of synthesis and transport of protein. **Smooth ER** has no ribosomes attached to it and is the site of lipid and steroid synthesis.

(d) Ribosomes

These are composed of ribosomal RNA and protein and are either bound to the rough RNA or free in the cytosol. They operate in conjunction with messenger RNA (mRNA) to assemble amino acids into protein. Often several ribosomes, collectively called a **polysome**, are associated with mRNA.

(e) Mitochondria

This organelle is bounded by two membranes. The inner one is folded into **cristae**. ATP is produced in mitochondria during aerobic respiration. The fluid **matrix** is the site of Krebs' cycle, while the cristae are involved in oxidative phosphorylation and electron transport (see the diagram in the answer to Worked Example 6.5).

(f) Golgi Apparatus

This consists of a stack of flattened cisternae and vesicles. It transports and chemically modifies materials within it, e.g. glycosylation occurs when carbohydrate is added to protein to form glycoprotein. Golgi apparatus may secrete materials by reverse pinocytosis, e.g. pancreatic enzymes, or package them into lysosomes.

(g) Lysosomes

These are small sacs containing **hydrolytic enzymes**. The enzymes are used to digest material taken in by endocytosis, autophagy (where unwanted structures in the cell are removed), exocytosis (for use within the organism), and autolysis during differentiation.

(h) Peroxisomes

These are bounded by a single membrane and contain **catalase**. They are concerned with metabolic activities involving oxidation. In plants, glyoxysomes convert lipids to sucrose in some seeds, and leaf peroxisomes are important in photorespiration.

(i) Microtubules

These are long, thin cylindrical organelles composed of the globular protein **tubulin**. They promote movement of materials within the cell and form a **cytoskeleton** which determines and maintains the shape of the cell. They are used in **centriole** formation (though centrioles are absent in the cells of higher plants) and spindle formation during cell division. Each centriole is composed of nine triplets of microtubules. This is also the structure of basal bodies which are the templates for cilia and flagella formation. Cilia and flagella possess enzymes which break down ATP. This generates energy to slide the microtubules over each other, hence creating ciliary or flagellar movement.

3.3 Structures Characteristic of Plant Cells

(a) Cell Wall

A primary cell wall is deposited on either side of the **middle lamella**. It consists of myofibrils of cellulose embedded in a matrix of pectins and hemicelluloses. The matrix resists compression and shearing while the cellulose is of high tensile strength.

Extra layers of cellulose, arranged in a cross-ply pattern, may be added to form the secondary wall. Lignification of some cell walls, e.g. xylem vessels, further improves their tensile and compressional strength.

The cell wall provides mechanical and skeletal support and aids development of turgidity in cells. Pits are formed between adjacent cells where there is an absence of cellulose. Here a cytoplasmic strand, the **plasmodesma**, links the cytoplasm of the two cells. Substances are transported from cell to cell via plasmodesmata.

(b) Plastids

These are formed from **proplastids** in meristematic regions. **Leucoplasts** are colourless and adapted for food storage. **Chromoplasts** synthesize coloured pigments which attract animals to flowers and fruits and so help pollination and dispersal respectively. **Chloroplasts** contain chlorophyll and carotenoids and are the sites of photosynthesis (see the diagram in the answer to Worked Example 4.2(a)).

(c) Vacuoles

A mature plant cell possesses a large permanent vacuole bounded by a single membrane, the **tonoplast**. The vacuole contains cell sap which helps generate turgor pressure, permits storage of waste products and provides nutrients for the cytoplasm. It may contain hydrolytic enzymes which promote autolysis when the cell dies.

3.4 Plant Tissues

(a) Tissues Consisting of One Type of Cell

(i) *Parenchyma*

This consists of living unspecialized cells distributed in the cortex, pith and vascular tissues. They act as packing tissue, and when turgid provide support for the plant. The cells are potentially meristematic and may be used for storage of starch. Intercellular spaces between them permit gaseous exchange. Modified parenchyma exists as:

epidermis: covering the whole plant; the waxy cutivle reduces water loss;

guard cells: regulate transpiration;

mesophyll: the main photosynthetic tissue, containing numerous chloroplasts;

endodermis: a selective barrier to the movement of water and salts in roots;

pericycle: contributing to secondary growth and lateral root development.

(ii) *Collenchyma*

This is situated in the outer regions of the cortex in young stems and petioles. It has walls of unevenly thickened cellulose, especially in the corners of the cell, and provides support and mechanical strength (see the diagrams of the sections of collenchyma in the answer to Worked Example 3.10).

(iii) *Sclerenchyma*

This is composed of lignified **fibres** and **sclereids**, both of which provide support. Their distribution is related to the stresses to which different plant organs are subjected. Fibres are arranged in strands or sheets and extend longitudinally for long distances. The ends of the cells interlock, thus increasing their combined strength (see the diagrams of the sections of sclerenchyma in the answer to Worked Example 3.10). Sclereids are widely dispersed either singly or in small groups. They confer rigidity. All sclerenchyma cells are dead.

(b) Tissues Consisting of More Than One Type of Cell

Two tissues, xylem and phloem, are in this category. Both are conducting tissues and comprise the vascular tissue of plants.

(i) *Xylem*

This has two main functions, the conduction of water and mineral salts and support. It is composed of four cell types. All four cell types have a support function, but only the first three are conducting tissues:

tracheids: elongated cells with tapered ends; these provide strength in the same way that the sclerenchyma fibres do and also possess paired pits;

vessels: tubular cells, wider and shorter than tracheids; they are perforated at each end, which facilitates easy passage of water;

parenchyma and **fibres**: as previously described under sclerenchyma (see the diagram of the section of xylem in the answer to Worked Example 3.10).

Protoxylem is formed initially by the embryo. As it matures it becomes more lignified to form **metaxylem**. Metaxylem may show **scalariform**, **reticulated** or **pitted** lignification. Mature xylem is dead tissue.

(ii) *Phloem*

This translocates organic solutes throughout the plant. The conducting elements are enucleate **sieve tube cells**. Characteristically, they possess **sieve plates** which permit the flow of solutions from one cell to another. Closely associated with the sieve elements are **companion cells**. Phloem parenchyma and fibres are also present (see the diagram of the section of phloem in the answer to Worked Example 3.10).

3.5 Animal Tissues

The four major types are epithelial, connective, muscular and nervous tissues.

(a) Epithelial Tissues

This is arranged in single or multi-layered sheets. The lower layer of cells is attached to a **basement membrane**. The main function of epithelium is to cover other tissues and protect them from injury through abrasion, pressure or infection. The free surface of epithelium is often highly modified to carry out absorptive, secretory, excretory or sensory functions. Epithelia are classified according to the number of cell layers, the shape of the cells and their cellular composition.

(i) *Simple Epithelia (One Cell Thick)*

1. **Squamous**: thin, flattened cells which permit efficient diffusion and smooth passage of fluids, e.g. endothelium, alveoli (see answer to Worked Example 3.8(b)(i)).
2. **Cubical**: cube-shaped cells lining many ducts. They may also be secretory, e.g. salivary glands.
3. **Columnar**: tall, narrow cells. May contain secretory goblet cells and microvilli, e.g. cells lining stomach and intestine.
4. **Ciliated**: columnar cells with cilia at their free surface. Contain secretory goblet cells. Cilia move materials and mucus from one region to another, e.g. oviducts and trachea (see answer to Worked Example 3.8(b)(ii)).

(ii) *Compound Epithelia (More Than One Layer Thick)*

Stratified: Cells near basement membrane are cubical or columnar. As they are pushed outwards they become flattened squames and protect underlying tissues from mechanical damage. Keratin may be added, making the squames cornified, e.g. vagina and external skin surfaces.

(iii) *Glandular Tissue*

Exocrine glands secrete their products into ducts. **Endocrine** glands pass secretions directly into the bloodstream and may be alternatively called **ductless** glands. **Mucus** cells secrete mucus materials, while **serous** cells secrete a solution containing enzymes. A mixed gland produces both types of secretion, e.g. pancreas.

(iv) *Connective Tissue*

This is composed of a fluid/semi-fluid matrix containing a variety of cells and fibres. It is the major supporting tissue of the body and some of its other functions include packing and binding of structures, protection against microbial invasion, insulation, and production of blood cells. There are several types of connective tissue:

1. **Areolar**: found around all organs in the body. It consists of a semi-fluid matrix containing a variety of cells and fibres. The cells include **fibroblasts** which make the fibres, **macrophages**, matrix-secreting **mast cells** and plasma cells which produce antibodies.
2. **White fibrous**: organized wavy bundles of collagen fibres which are strong, flexible yet inextensible. Abundant in tendons and ligaments.
3. **Yellow elastic**: loose network of fibres. Present in ligaments and artery walls.
4. **Adipose**: areolar tissue containing many fat cells which act as an energy reserve, for insulation and as a shock absorber.
5. **Haemopoietic tissue**: this forms red and white blood cells and is located in the red bone marrow and lymphoid tissue of mammals.

(v) *Skeletal Connective Tissue*

1. **Cartilage**: the **chondroblast** cells are embedded in a resilient **matrix** of **chondrin** and enclosed in spaces called **lacunae**. Cartilage is tough and flexible and can effectively resist stresses and strains. Hyaline cartilage is the only skeletal material of elasmobranchs and forms the embryonic skeleton in bony vertebrates.
2. **Bone**: This is calcified connective tissue. Bone-forming **osteoblasts** are contained in **lacunae** throughout the **matrix**. Resorption and reconstruction processes enable a bone to adapt its structure to changing mechanical requirements. Two hormones, calcitonin and parathormone, control the release of calcium and phosphate. There are three types of bone:
   (a) **Compact bone**: This consists of concentric cylinders of bony **lamellae** surrounding a central **Haversian canal** which contains an artery and vein. Osteoblasts interspersed between the lamellae aid bone deposition; inactive osteoblasts are called **osteocytes**. A layer of dense connective tissue, the **periosteum**, covers the surface of the bone.
   (b) **Spongy bone**: a meshwork of thin bony **trabeculae**. Marrow fills the spaces between the trabeculae.
   (c) **Membrane bone**: groups of osteoblasts manufacture flat bony plates close to the surface of the body, e.g. the bones of the cranium.

(vi) *Muscle Tissue*

This is composed of specialized contractile cells or fibres held together by connective tissue. Three types exist:

1. **Skeletal**: attached to the skeleton. It comprises long cells containing a large number of mitochondria, a prominent smooth ER and T-system. It is generally

under voluntary nervous control. It produces powerful, rapid contractions but fatigues quickly.

2. **Involuntary**: often found surrounding hollow structures such as blood vessels, the gut and ducts. It possesses spindle-shaped cells and is under the control of the autonomic nervous system. It produces slow, sustained contractions but does not fatigue.

3. **Cardiac**: only found in the heart. The cells are branched and connected to each other by intercalated discs. It is myogenic and contracts powerfully, but does not fatigue.

(vii) *Nervous Tissue*

Composed of **neurones** specialized for the conduction of nerve impulses, and accessory **neuroglial cells**. Neurones provide a means of rapid communication between receptors and effectors. *Sensory* (afferent) neurones have a different structure from *motor* (efferent) neurones associated with their positions within the nervous system. Afferent neurones conduct impulses to the central nervous system where internuncial neurones pass the impulses via synapses to efferent neurones. Each neurone has a **cell body** and a variable number of cytoplasmic processes extending from it. **Dendrons** conduct impulses towards the cell body while **axons** conduct impulses away from the cell body. **Nerves** consist of bundles of neurones. Cranial and spinal nerves are **myelinated**. Nerves of the autonomic nervous system are non-myelinated.

3.6   Worked Examples

3.1

Draw a large labelled diagram to show the ultrastructure of a generalized animal cell.   [8]

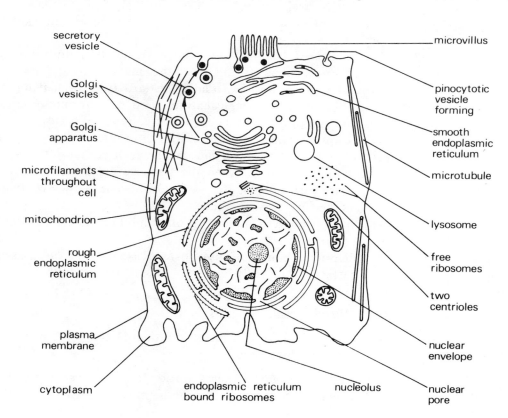

3.2

List four differences between plant cells and animal cells. [4]

Plant cells	Animal cells
(a) cellulose cell wall with plasmodesmata	no cell wall or plasmodesmata
(b) large, permanent, central vacuole	many, much smaller, temporary vacuoles
(c) centriole absent (in higher plants)	centrioles present
(d) microvilli absent	microvilli present

(Note to reader: There are many other possible correct answers. Only four examples, in each case, have been given here.)

3.3

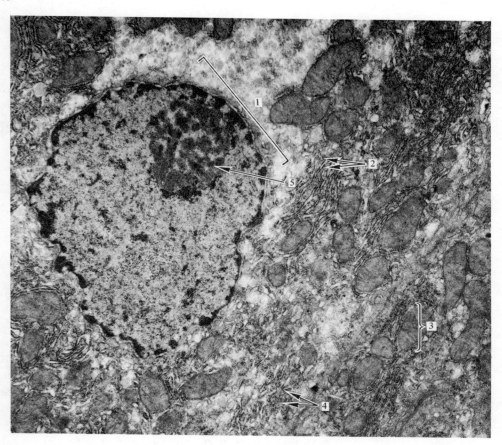

Fig. 3.1

Figure 3.1 is an electron micrograph of part of a liver cell.

(a) Name the parts 1 to 5 indicated on the photograph.

1 nuclear membrane   4 ribosomes

2 rough endoplasmic reticulum 5 nucleolus

3 mitochondrion   [4]

(b) *If the magnification of the photograph is 20 000 times, what is the actual length of structure 3?*

Structure 3 measures 2 cm

$\dfrac{2}{20\,000} = 1.0 \times 10^{-6}$ m. [2]

(c) *State the functional sequence connecting structures 1, 2, 3, 4 and 5.*

mRNA leaves nucleus through 1. Attaches to ribosomes at 2. Protein synthesis occurs at 2 using energy produced by 3. rRNA found in 2 and 4 is produced by 5.

[3]

(d) *Structure 3 has folds inside it. What is their significance?*

These are called cristae. Enzymes involved in aerobic respiration are situated within the cristae which have a large surface area. [2]

[L]

3.4

Give two features by which you could distinguish a prokaryotic cell from a eukaryotic cell.

A prokaryotic cell has no nucleus or any other membranous organelle and no microtubules. A eukaryotic cell has all these features.

[2]

[OLE]

3.5

(a) *How has cellular ultrastructure been studied?*

A detailed knowledge of cellular ultrastructure has come from evidence revealed by electron microscopy. A steady stream of electrons is emitted from the cathode as a result of a high voltage applied between the cathode and the anode. The electrons are focused by electromagnets as they pass through the high vacuum in the microscope tube. The electrons pass through the specimen and the scattered electron pattern which shows cellular ultrastructure is seen as a visible image on a fluorescent screen. Specimens must be cut very thinly using an ultramicrotome and are usually stained with heavy metal compounds such as lead nitrate, or heavy metals such as gold or platinum.

The scanning electron microscope sends a focused beam of electrons to and fro across the surface of a specimen and the electrons reflected from the surface are focused onto a screen. A three-dimensional effect is produced which reveals great detail but not to the same extent as with the transmission electron microscope described above.

The resolving power of these microscopes is about 500 times greater than that produced by a light microscope.

(b) *What problems of interpretation of cell structure have been encountered?* [OLE]

Because of the complex nature of the fixation, embedding and staining of specimens for study in the electron microscope and the fact that the material is no longer living it is possible for structures to become damaged and displaced. Many specimens must be studied to eliminate such possible sources of error. The various staining procedures, too, may produce shadows and distortions and give a false impression of ultrastructure. Such features are described as artefacts.

3.6

Describe the roles of each of the following membranous organelles: Golgi apparatus, lysosomes, microtubules, plastids, endoplasmic reticulum. [12]

[UCLES] [part question]

The information required to answer this question is given in sections 3.2 and 3.3.

3.7

(a) *Give two functions of cell membranes.* ...They control the exchange of materials between the cell and the environment and enable separate compartments to be formed within cells.

.... [2]

(b) (i) *Give the principal constituents of cell membranes.* ...proteins and phospholipids
[2]

(ii) *Draw a diagram to illustrate the arrangement of these constituents in the fluid mosaic model of the cell membrane.*

(iii) *Indicate on your diagram a cell-surface receptor (a protein involved in the binding of hormones and other substances), and label the external and cytoplasmic faces of the membrane.* [6]

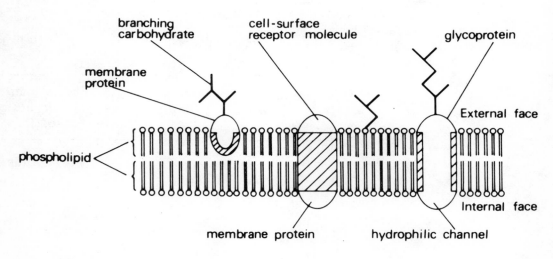

39

3.8

(a) *Explain what is meant by a tissue in a living organism.*

A collection of structurally similar cells specialized to perform a particular function or functions.

.... [4]

(b) *For both (i) pavement (squamous) epithelium and (ii) ciliated epithelium, state two locations in the body of a mammal and the importance of the tissue in these locations.*

(i) *Pavement (squamous) epithelium*

first location endothelial lining of blood vessels

importance to provide a smooth lining for blood flow with minimum turbulence

second location wall of alveoli

importance to provide a thin layer for rapid diffusion of gases

.... [4]

(ii) *Ciliated epithelium*

first location oviduct wall

importance to propel the oocyte along the oviduct towards the uterus

second location tracheal lining

importance upward movement of mucus containing foreign particles towards the throat

.... [4]

(c) *In the table below list four structural differences (i–iv) between the epidermis of a mammal and the epidermis of the leaf of a dicotyledonous plant. Write each part of your answer within the spaces provided in the table.*

	Epidermis of a mammal	Leaf epidermis
(i)	no stomata present	may contain stomata
(ii)	no cuticle present	cuticle present
(iii)	hairs pass through from hair follicles	'hairs' may be present but do not arise in follicles
(iv)	Malpighian layer actively dividing	no Malpighian layer

[4]
[L]

40

3.9

The sieve tube is an important cell element in the phloem of flowering plants.

(a) List two structural features characteristic of sieve tubes:

(i) *have cellulose cell walls* ....

(ii) *sieve plates present* ....

[2]

(b) Name one other type of cell which occurs in phloem tissue.

companion cell .... [1]

(c) How does this cell differ from a sieve tube?

has a nucleus .... [1]

[OLE]

3.10

By means of labelled annotated diagrams show the differences between the following pairs of cells and tissues:

    sclerenchyma and collenchyma   [4, 4]

    phloem and xylem   [6, 6]

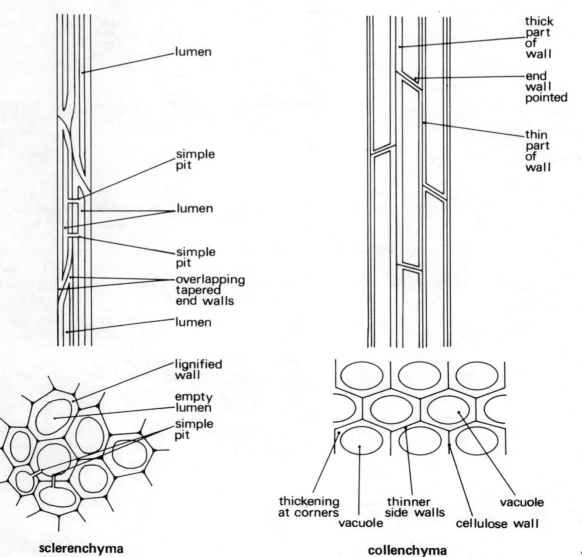

sclerenchyma   **collenchyma**

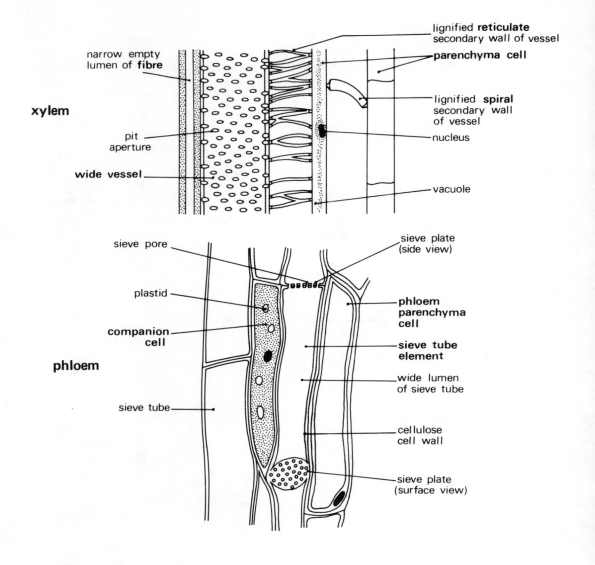

xylem

narrow empty
lumen of **fibre**

lignified **reticulate**
secondary wall of vessel

parenchyma cell

lignified **spiral**
secondary wall
of vessel

pit
aperture

nucleus

wide vessel

vacuole

phloem

sieve pore

sieve plate
(side view)

plastid

**phloem
parenchyma
cell**

companion
cell

**sieve tube
element**

wide lumen
of sieve tube

sieve tube

cellulose
cell wall

sieve plate
(surface view)

3.11

Describe, with reference to skeletal connective tissue, the relationship between structure and function. [10]

Cartilage is a tough yet flexible tissue which has to resist stresses and strains. The basic structure of cartilage, as shown in Fig. 3.2, is seen to consist of an elastic, compressible matrix called hyaline produced by cells added chondroblasts. This

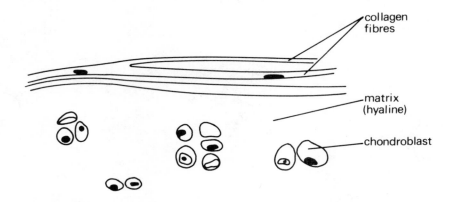

collagen
fibres

matrix
(hyaline)

chondroblast

Fig. 3.2

type of cartilage is found whenever rigidity is required, e.g. in the nose and at the ends of bones. Other materials can be incorporated in this basic structure to produce other types of cartilage with different functions. For example, cartilage containing yellow elastic fibres has greater elasticity and flexibility as in the external ear, epiglottis and cartilages of the pharynx, while cartilage containing bundles of white collagen fibres provides greater tensile strength; this white fibrous cartilage is found as discs between adjacent vertebrae in the vertebral column where it has a cushioning effect.

Bone, as shown in Fig. 3.3, consists of cells called osteoblasts embedded in a matrix composed of hydroxyapatite and collagen fibres. This structure can withstand compression strains and resist tension. Because of the relationship between blood vessels and osteoblasts the structure of bone can be adapted to carry out varying mechanical functions including support and protection. In the embryo and in the limb bones the bone needs to be softer. This type is called spongy bone. The matrix contains less inorganic material, therefore is lighter and less brittle than the harder, compact bone which forms the bulk of the skeleton in adults. There is a third type of bone called membrane bone, which is even harder than compact bone. It forms the flat, plate-like components of the skull, jaws and pectoral girdle.

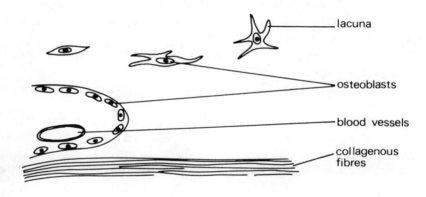

Fig. 3.3

Dentine is the hardest skeletal material of all. It has an structure entirely different from that of bone. However, it is composed of the same basic materials but has a higher inorganic content. This makes it much harder. There are many spaces within dentine occupied by blood vessels and nerve endings and this structure is highly adapted to its function in resisting compression and providing a tough, resistant material for chewing, biting and tearing of food.

4 Autotrophic Nutrition

Phototroph   An organism which utilizes light energy in order to synthesize organic molecules. (Greek: *photos*, light.)

Autotroph   An organism which utilizes an inorganic source of carbon for purposes of its nutrition.

Heterotroph   An organism which utilizes an organic source of carbon for purposes of its nutrition.

Holophytic   (sometimes used in place of autotrophic) A form of nutrition which utilizes light energy and an inorganic source of carbon to synthesize organic molecules.

Limiting factor   In a chemical reaction involving several factors, the factor which prevents the reaction rate from increasing because it is present at its minimum value is called the limiting factor.

Compensation point   The point at which the rate of uptake of carbon dioxide is equal to the rate of output of carbon dioxide.

Action spectrum   The relationship between rate of photosynthesis and wavelength of light.

Absorption spectrum   The relationship between absorbance of light and wavelength of light, for photosynthetic pigments.

4.1   Autotrophism

Autotrophs synthesize **organic** materials from **inorganic** materials. Some organisms derive their energy for this process from light and are called **photoautotrophs**, e.g. green plants. Other organisms use chemical energy and are called **chemoautotrophs**, e.g. nitrifying bacteria. Photoautotrophs are the **primary producers** of food chains.

4.2   Photosynthesis

Photosynthesis is a multi-stage process consisting of a series of light-requiring reactions and a series of chemical reactions. The former, collectively called the **light reaction**, involves **light energy**, absorbed by chlorophylls and carotenoids. This light energy 'splits' water molecules (**photolysis**) into hydrogen and oxygen. The latter is released as **gaseous oxygen**. In the non-light-dependent reactions, or **dark reactions**, hydrogen combines with (reduces) **carbon dioxide** to form **carbohydrate**. The overall equation can be represented as shown at the top of the next page.

$$CO_2 + 2H_2O \xrightarrow[\substack{chlorophyll \\ + \\ enzymes}]{\substack{light \\ energy}} [CH_2O] + O_2\uparrow + H_2O$$

4.3 Leaf Structure

For efficient photosynthesis the leaf should be thin, have a large surface area for absorption of light and gaseous diffusion, and a means of preventing excessive water loss through **stomata** and **epidermis**. Large numbers of **chloroplasts** in **palisade mesophyll** cells provide the main photosynthetic tissue. The spaces between the irregularly shaped spongy mesophyll cells within the leaf permit free **diffusion** of gases. **Turgor** changes in the **guard cells** permit gaseous exchange with the atmosphere. **Cuticle** on the single-layered **transparent** upper and lower epidermis protects the leaf from desiccation and infection.

4.4 Chloroplasts

These biconvex organelles, containing many flattened, fluid-filled membranous sacs called **thylakoids** and a gel-like **stroma**, are enclosed by the two membranes of the **chloroplast envelope**. Stacks of circular thylakoids called **grana**, linked together by **intergranal lamellae**, are formed at intervals throughout the chloroplast.

4.5 Photosynthetic Pigments and Light

Molecules of **chlorophyll a**, **chlorophyll b**, **carotene** and **xanthophyll** are situated in the thylakoid membranes.

The **absorption spectrum** for isolated pigments shows the percentage of light absorbed by each pigment at different wavelengths. The **action spectrum** shows how effective these pigments are in stimulating photosynthesis. Pigments absorb light energy in the **blue** and **red regions** of the **visible spectrum**. The light energy causes excitation of electrons in the outer shells of magnesium atoms of the chlorophyll molecules. These electrons are displaced to **higher energy levels** or 'excited states'. The latter are unstable. Electron transfer occurs as electrons from **electron donor** chlorophyll molecules fall back towards the **ground state**. A chain of **electron carrier** molecules, including the iron-containing protein **ferredoxin**, passes electrons to **electron acceptor** molecules. The main acceptor molecule is **nicotinamide–adenine dinucleotide phosphate (NADP)**.

4.6 'Light Reaction'

Excited **accessory pigments** pass emitted electrons to two chlorophyll a **primary pigments** called P700 and P690. Each primary pigment forms the **reaction centre** of two photosynthetic units called **photosystems I and II (PSI and PSII)** respectively. These can be seen as structures called **quantasomes** in the thylakoid membranes. As 'high-energy' electrons from PSII flow along the electron transport chain to PSI the energy given out is coupled to the formation of **ATP**. This is called **non-cyclic photophosphorylation**.

'High-energy' electrons from PSI then pass to NADP where they combine with hydrogen ions (H^+), which come from the initial photolysis of water, to form **NADPH$_2$**. Electrons released from water during this photolysis replace those electrons lost from PSII.

Gaseous oxygen is released and ATP and $NADPH_2$ are used in the 'dark reactions'.

4.7 'Dark Reaction'

Carbon dioxide, in aqueous solution, diffuses into the stroma where it reacts with the 5C **carbon dioxide acceptor, ribulose bisphosphate (RuBP)**. This reaction is catalysed by the enzyme **RuBP carboxylase**. The product, a 6C unstable compound, immediately breaks down to form two molecules of 3C **phosphoglyceric acid (PGA)**. Hydrogen atoms from $NADPH_2$ reduce PGA to a 3C sugar, **triose phosphate (TP)**, using energy from the breakdown of ATP.

Of the TP formed one-sixth is converted into **hexose sugars** such as glucose. Hexose sugars may be further converted into sucrose, starch, amino acids or fatty acids. The remainder of the TP is used to regenerate RuBP by a series of reactions called the **Calvin cycle**.

4.8 Limiting Factors

Temperature, carbon dioxide concentration and **light intensity** are three external factors which may limit the rate of photosynthesis. The initial relationship between rate of photosynthesis and any factor is linear. At a certain point the reaction rate will stabilize due to one or more factors being in short supply. At this point the factor is said to be limiting, as shown in Fig. 4.1.

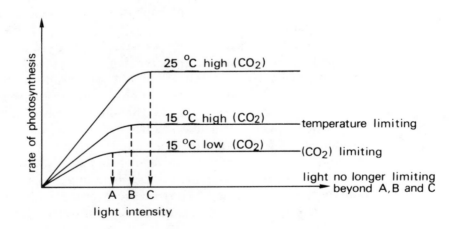

Fig. 4.1

4.9 C$_4$ Plants

Many plants, including maize and sugar-cane, are far more efficient at taking up CO_2 than C_3 plants. The CO_2 acceptor in C_4 plants is **phosphoenolpyruvate (PEP)**. PEP reacts with CO_2 to form oxaloacetic acid which is reduced by $NADPH_2$ to form malic acid. The malic acid then reacts with RuBP to form **pyruvic acid** and PGA. The pyruvic acid is then phosphorylated by ATP to regenerate PEP while PGA is converted to TP as for C_3 plants. These reactions are called the **Hatch–Slack pathway**.

C_4 plants have two rings of cells surrounding the vascular bundles. The outer **mesophyll cells** fix CO_2 using PEP and pass malic acid to the inner **bundle sheath cells**. Here CO_2 is released for acceptance by RuBP and the eventual formation of TP. The photosynthetic efficiency of C_4 plants is due to their ability to fix CO_2 in environmental conditions where CO_2 is the limiting factor.

4.10 Compensation Point

This is the point at which the rate of uptake of carbon dioxide by plants (for photosynthesis) is balanced by the rate of output of carbon dioxide (from respiration). If light intensity and carbon dioxide concentration are not limiting factors, the time taken to reach this point, from darkness, is called the **compensation period**.

4.11 Experimental Investigations

It is essential that you are familiar with full experimental details of the following experiments and techniques:

1. Destarching a plant.
2. Testing a leaf for starch.
3. Investigating the need for light.
4. Investigating the need for carbon dioxide.
5. Investigating the need for chlorophyll.
6. Investigating the evolution of oxygen.
7. Investigating gaseous exchange in leaves.
8. Investigating the effect of light intensity on the rate of photosynthesis.

4.12 Chemoautotrophs

These organisms are bacteria using CO_2 as a carbon source and energy derived from the oxidation of inorganic materials such as iron, sulphur, ammonia and nitrite. These bacteria are particularly important in **nitrogen fixation**, **nitrification** and **denitrification**.

4.13 Mineral Nutrition

Successful growth and development requires adequate supplies of **micronutrients (trace elements)** and **macronutrients**. Shortage of any of these nutrients can lead to **deficiency diseases**, such as **chlorosis**.

4.14 Worked Examples

4.1

*Read through the following account of photosynthesis and then write on the dotted lines the most appropriate word or words to complete the account. (In this answer the missing words are shown in **bold** type.)*

*There are four pigments commonly found in the chloroplasts of higher plants; chlorophyll a, chlorophyll b, carotene and **xanthophyll**. Chlorophyll a absorbs mainly red and **blue** light. The absorption of light causes the displacement of an electron from the chlorophyll a molecule. This electron may be passed back to the chlorophyll via a series of **electron carriers** which are at a progressively lower energy level. Coupled with this electron transfer is the synthesis of **ATP**. This compound may be subsequently used in the dark reaction of photosynthesis which occurs in the **stroma** region of the chloroplast. During non-cyclic photophosphorylation, the electron is combined with **hydrogen** ions resulting from the photolysis of **water** to form the reduced coenzyme called **NADPH$_2$**. This reduced coenzyme is used in the **Calvin** cycle to convert **phosphoglyceric** acid to phosphoglyceraldehyde, which can be converted to **ribulose bisphosphate** which is the acceptor molecule for the carbon dioxide used in photosynthesis. The electron emitted from the chlorophyll molecule is replaced by electrons from the **hydroxyl** ions produced by the photolysis reaction. As a result **oxygen** gas is given off.* [14]

[L]

4.2

(a) In the space below, make a labelled diagram of a chloroplast to show its structure as seen under an electron microscope. [6]

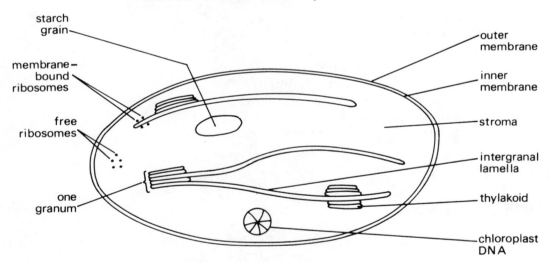

(b) Name, and give the colour of, the four pigments that usually make up the 'chlorophyll' in a leaf. [4]

Name of pigment	Colour
(i) chlorophyll a	yellow-green
(ii) chlorophyll b	blue-green
(iii) carotene	orange
(iv) xanthophyll	yellow

(c) *What do you understand by the term 'action spectrum'?*   [2]
                                                                               [L]

A graphical representation of the relationship between rate of photosynthesis and wavelength of light.

4.3

Photosynthesis occurs in two stages, a light-dependent stage ('light reaction') and a light-independent stage ('dark-reaction').

(a) *Where precisely in the plant cell does each stage occur?*
   (i) the light-dependent stage
       In the membrane of the thylakoid and grana.
  (ii) the light-independent stage
       In the stroma of the chloroplast.

(b) *(i) Which two compounds produced in the light-dependent stage are subsequently used in the light-independent stage?*
       Adenosine triphosphate (ATP) and $NADPH_2$.
  (ii) How are these two compounds produced?
       ATP is produced as electrons pass along an electron carrier chain during non-cyclic photophosphorylation. $NADPH_2$ is produced from oxidized NADP and $2H^+$ released from photolysis of water.

(c) *Which compound in C_3 plants acts as the acceptor of carbon dioxide in the light-independent stage?*
       Ribulose bisphosphate (RuBP).   [7]
                                                                            [AEB]

4.4

Graph A (Fig. 4.2) shows the absorption spectrum of a solution of pigments extracted from bean leaves and the action spectrum of a bean plant determined by measurement of the rate of photosynthesis when illuminated by different wavelengths of light.

Graph B shows the absorption spectra of three pigments which have been extracted from bean leaves and examined individually.

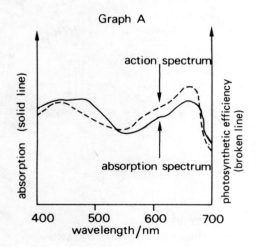
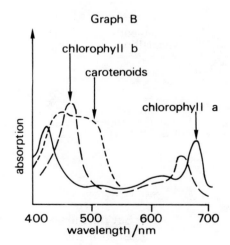

Fig. 4.2

(a) With reference to graph A,

    (i) comment on the biological significance of the relationship between the action spectrum and the absorption spectrum;

The close similarity between the two curves shows that the wavelengths absorbed by the pigments lead to the most efficient rates of photosynthesis by the bean leaves.

    (ii) relate the visible colour of a leaf to the absorption spectrum;

The leaf is green because wavelengths in this region of the spectrum, 550 nm, are less strongly absorbed by the leaf. These wavelengths are reflected.

    (iii) state the approximate wavelength at which photosynthesis is most efficient.   **[4]**

660 nm

(b) With reference to graphs A and B account for the difference between photosynthetic rate and light absorption at 490 nm.   **[2]**

The absorption spectrum at 490 nm in graph B shows that carotene is the pigment absorbing light. The action spectrum in graph A shows that photosynthetic efficiency is relatively lower here than at any point in the spectrum. Carotene is, therefore, not so efficient at using this light energy for photosynthesis. It, in fact, passes the energy on to chlorophyll b or a.

(c) Describe how you would carry out an experiment to determine the action spectrum for photosynthesis.   **[6]**

Experiment to investigate the efficiency of photosynthesis over a range of wavelengths of light from 700 nm (red) to 400 nm (blue).

APPARATUS

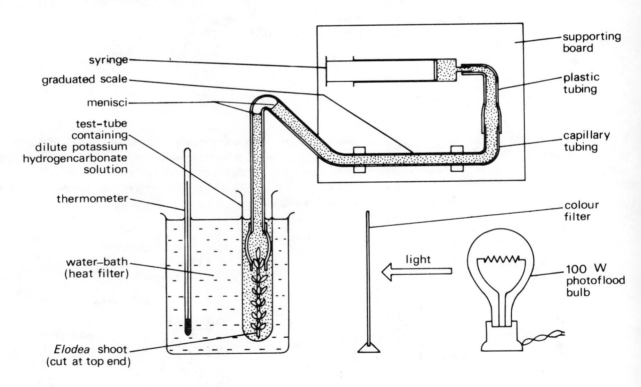

METHOD

1. Set up the apparatus as shown in the diagram above.
2. Ensure that the only variable throughout the experiment is the wavelength of light, by:
   (a) keeping the temperature of the water-bath constant,

(b) providing an excess of carbon dioxide in the test-tube (supplied by the potassium hydrogencarbonate solution),

(c) keeping the light intensity constant by varying the distance of the light source from the plant.

3. Leave the apparatus without a filter for 15 min to equilibrate.

4. During this time, familiarize yourself with the technique of withdrawing the bubbles of gas collected and measuring their total volume in the capillary tubing. The volume of gas evolved should be measured for each 5-min period.

5. Insert the blue filter and allow the apparatus to equilibrate for 10 min.

6. Record the volume of gas evolved during each 5-min period for a total of 15 min.

7. Repeat 5 and 6 using each of the following coloured filters in turn, in place of the blue filter: green, yellow, orange, red.

8. Tabulate your results.

9. Calculate the mean volume of gas evolved during 5 min for each colour filter.

10. Plot a graph using wavelength of light/nm as the x-axis and volume of gas evolved/cm as the y-axis. This graph is an action spectrum for wavelengths of light between 400 and 700 nm.

(d) Construct a diagram to show the roles of chlorophyll a in non-cyclic photophosphorylation and photolysis. [5]

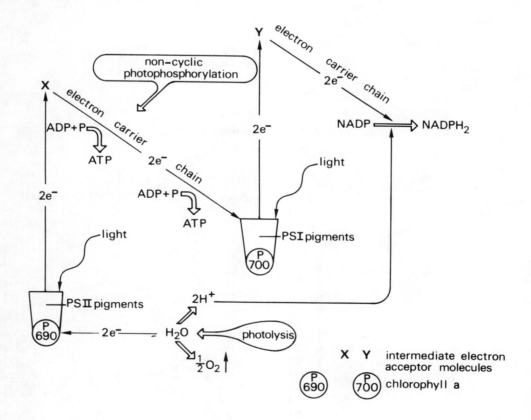

(e) Non-cyclic photophosphorylation occurs in the grana of chloroplasts. State precisely where oxidative phosphorylation occurs. [3]

[AEB]

Oxidative phosphorylation occurs at various points between hydrogen acceptor molecules and cytochrome molecules of the electron transport chain found on the membranes of the cristae in mitochondria.

Plant species A and B grow naturally in different habitats. In an experiment the exchange of carbon dioxide between the atmosphere and species A and B was determined over a range of light intensities from darkness to the equivalent of mean noon sunlight. A constant temperature was maintained throughout the experiment. The amount of carbon dioxide absorbed or released was determined by measuring the carbon dioxide concentration in a stream of air before and after it had passed over the plants. The data obtained are given below.

Light intensity as a percentage of mean noon sunlight	Net carbon dioxide absorption in arbitrary units	
	Species A	Species B
0	−0.1	−0.8
10	+3.0	+0.5
20	+5.3	+3.5
30	+6.5	+7.0
40	+6.5	+9.3
50	+6.7	+11.5
60	+6.8	+13.2
70	+7.0	+15.0
80	+6.5	+17.0
90	+6.8	+18.0
100	+6.7	+19.0

(a) Plot these data on a single set of axes. [5]

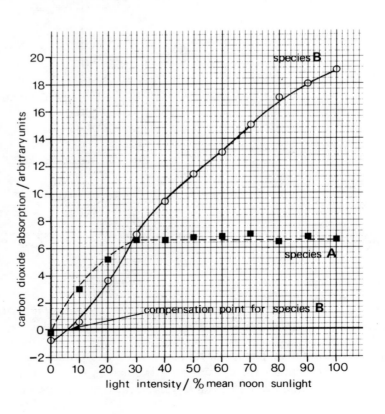

52

(b) Discuss the extent to which species A and species B might be able to grow in the same habitat. [4]

The amounts of carbon dioxide taken up by species A and B indicate their rates of photosynthesis.

The graph shows that the compensation point for species A is reached at a light intensity of 1 per cent, whereas the compensation point for species B occurs at a corresponding value of 6 per cent. This indicates that species A is able to carry out photosynthesis faster at lower light intensities than species B (i.e. up to a light intensity of 30 per cent of mean noon sunlight). Beyond light intensities of 30 per cent mean noon sunlight, species B has a much higher rate of photosynthesis. From the data given, species A would appear to be a 'shade species' and thus be physiologically adapted to make efficient use of low light intensities. Species B would appear to be a 'sun species' and adapted to make efficient use of high light intensities. Species A is unable to increase its rate of photosynthesis even if the light intensity increases beyond 30 per cent. The above information suggests that these plants would be entirely capable of growing in the same habitat. However, competition for light may act as a limiting factor. If the leaves of species A are situated above those of species B this would present problems for growth of species B, which depends upon higher light intensities. This would not be a problem if the leaves of species B were situated above those of species A.

(c) (i) What is meant by the term 'compensation point'?

The compensation point is the point at which the rate of uptake of carbon dioxide by a plant is equal to the rate of output of carbon dioxide. It indicates the point at which the rate of photosynthesis and rate of respiration are exactly equal.

(ii) Clearly indicate on your graph the compensation point for species B. [2]

See label on graph in part (a).

(d) (i) What is meant by the term 'limiting factor'?

A limiting factor is the factor, in a chemical reaction involving several factors, which prevents the reaction rate from increasing because it is present at its minimum value.

(ii) From your knowledge of photosynthetic pathways, explain precisely how three named factors can be limiting in photosynthesis. [7]

1. Temperature

This influences the rate of the enzyme-catalysed reactions of the light-independent stages of photosynthesis. At low temperatures, the activity of ribulose bisphosphate carboxylase will limit the uptake of carbon dioxide by ribulose bisphosphate. Thus the rate of formation of phosphoglyceric acid and triose phosphate will be reduced.

2. Light intensity

This can limit the initial light-requiring reactions of photosynthesis. Inadequate light intensity will not provide sufficient energy to excite electrons in the magnesium atoms of the chlorophyll molecules to higher energy levels. No electrons will flow along the electron carrier chain and non-cyclic photophosphorylation will not occur. No ATP will be produced. No photolysis of water will occur to replace the electrons lost from chlorophyll and no $NADPH_2$ will be formed.

3. Carbon dioxide concentration

If a plant has optimum light intensity and optimum temperature, the limiting factor of photosynthesis will be carbon dioxide concentration. Carbon dioxide is a reactant required in the initial chemical reactions of the light-independent Calvin cycle.

(e) Distinguish between C_3 and C_4 plants. [2]

[AEB]

A C_3 plant is so called because the initial photosynthetic product is a 3C acid, phosphoglyceric acid, as opposed to the 4C acid, oxaloacetic acid, as in C_4 plants. C_3 plant leaves contain only one type of chloroplast, whereas C_4 plants have two types of chloroplast, one in the mesophyll cells and another in the bundle sheath cells. In C_3 plants the carbon dioxide acceptor is ribulose bisphosphate, while in C_4 plants it is phosphoenolpyruvate. C_3 plants, e.g. potato, have a lower photosynthetic efficiency than C_4 plants, e.g. maize and sugar cane.

4.6

Which one of the following compounds is produced by the reaction between carbon dioxide and phosphoenolpyruvate (PEP) in tropical plants, such as sugar cane?
A *malic acid*
B *oxaloacetic acid*
C *phosphoglyceric acid*
D *ribulose bisphosphate*
E *triose phosphate*

The correct answer is B, which is 'the product of the reaction between CO_2 and PEP in C_4 plants'. The distractors are:

A is a 4C acid formed from B.
C is the initial product in C_3 plants.
D is the carbon dioxide acceptor in C_3 plants.
E is an initial sugar formed during photosynthesis.

4.7

The production of starch by leaves was investigated by removing discs from a destarched leaf. The discs were divided between four flasks labelled A to D and treated as shown below. Each flask was kept in air. After 36 hours the discs were tested for the presence of starch.

Flask	Fluid contents of flasks	Conditions in which flasks were kept	Results of testing for starch
A	Water enriched with carbon dioxide and kept at 25 °C.	Light	Present
B		Dark	Absent
C	Glucose solution enriched with carbon dioxide and kept at 25 °C.	Light	Present
D		Dark	Present

(a) Explain how and why the leaf was destarched. [3]

Leave a leaf in darkness for 48 hours. During this time starch is either broken down by enzymes and transported out of the leaf or used up in respiration. If the presence of starch is to be demonstrated after treatment none should be initially present.

(b) *Explain why the water and glucose solutions were enriched with carbon dioxide.* [2]

Carbon dioxide is a raw material for photosynthesis. Excess carbon dioxide must be present so that it cannot be a limiting factor.

(c) *Explain the results shown in the four flasks A, B, C and D.* [8]

[L]

A Carbon dioxide, water, temperature and light enable photosynthesis to produce more carbohydrate than is used in respiration. Starch is stored.

B Light is not present, therefore there is no photosynthesis. Starch is used up in respiration.

C All requirements for photosynthesis are present. Presence of glucose enables even more starch to be formed.

D No light is present; all other requirements are available for photosynthesis. Glucose is used to provide energy for phosphorylation. Starch is formed.

4.8

Describe the process of nitrogen fixation by living organisms. How is this nitrogen made available to other green plants? [10]

Gaseous nitrogen can be converted into a form which can be utilized by plants by the action of chemoautotrophic organisms in two ways. Free-living nitrogen-fixing bacteria in the soil belonging to the genera *Azotobacter* and *Clostridium* and several types of blue–green algae, e.g. *Nostoc*, take up gaseous nitrogen from the soil atmosphere and reduce it to form ammonia:

$$N_2 + 3H_2 \rightleftharpoons \underset{\text{ammonia}}{2NH_3}$$

Symbiotic bacteria of the genus *Rhizobium* enter the plant roots of leguminous plants, e.g. beans, peas and clover, and cause proliferation of root cortex cells to form swellings called root nodules. Inside these nodules the bacteria carry out the above reaction. The ammonia, as ammonium ions (NH_4^-), is used by the plants to make amino acids and nucleic acids.

When dead plant materials, including roots and root nodules sloughed off during growth, are decomposed by the action of saprophytic bacteria and fungi, the proteins, amino acids and nucleic acids are broken down into ammonium ions in the soil solution. In well-aerated soils this ammonia is oxidized by nitrifying bacteria to form nitrates which can be absorbed by the root system of all green plants. These oxidation reactions, and bacteria involved in them, are:

$$\underset{\text{ammonia}}{NH_3} \xrightarrow{\textit{Nitrosomonas}} \underset{\text{nitrite}}{NO_2^-} \xrightarrow{\textit{Nitrobacter}} \underset{\text{nitrate}}{NO_3^-}$$

4.9

The following questions relate to plant nutrition.

(a) *State what is meant by, and give two examples each of,*

(i) *a trace element;*

an element, used in low concentrations, required for healthy growth

first example:
zinc
second example:
copper

(ii) *a deficiency symptom.*
a visible effect of lack of an essential mineral ion
first example:
chlorosis in leaves (yellowing) due to lack of nitrogen
second example:
yellowing of leaf margins due to lack of potassium [6]

(b) *Give two uses to the plant of the following ions:*

(i) *nitrate;*
first use:
formation of amino acids, e.g. valine
second use:
formation of nucleic acids, e.g. RNA

(ii) *potassium;*
first use:
enzyme activator
second use:
maintains electrolyte balance in cell vacuole

(iii) *phosphate;*
first use:
synthesis of nucleic acids
second use:
formation of cell membrane (phospholipids)

(iv) *magnesium.*
first use:
formation of chlorophyll
second use:
co-factor in enzymes (phosphatases) [8]
[L]

5 Heterotrophic Nutrition

> **Digestion** The breakdown, by physical or chemical means, of large complex insoluble foodstuffs into small simple soluble molecules.
>
> **Absorption** The uptake of simple soluble molecules into living cells.
>
> **Peristalsis** Waves of muscular contraction passing along tubular organs, e.g. the gut.
>
> **Saprophyte** An organism that feeds on dead and decaying organic matter.
>
> **Symbiont** An organism which forms a close association with an organism of another species.
>
> **Parasite** An organism that lives in or on another living organism (called the host), generally receiving shelter and deriving nutrients from it. The parasite may cause harm to the host.
>
> **Heterotrophs** eat ready-made complex organic food. From this they obtain energy for metabolism, atoms and molecules to build new protoplasm or repair worn-out parts, and ions, coenzymes and vitamins vital for chemical processes. There are *four* types of heterotrophic nutrition.

5.1 Holozoic Nutrition

Organisms feed on solid organic matter. For the food to be made available to the body it must be in the form of simple molecules, small enough to pass through cell membranes into the body of the heterotroph. Animals feeding on plants are called **herbivores**; those that feed on other animals are called **carnivores**. Animals eating animal and vegetable matter are called **omnivores.** The processes involved in holozoic nutrition are:

1. **Ingestion**: the intake of food;
2. **Digestion**: the breakdown of the food into simple molecules;
3. **Absorption**: the uptake of these simple molecules into living cells;
4. **Assimilation**: the use to which the absorbed molecules are put;
5. **Egestion**: expulsion of undigested waste food materials from the body.

5.2 Digestion in Man

The gut of man is a continuous muscular tube running from mouth to anus. It is differentiated throughout its length, with each region specialized for a particular phase in the overall process of digestion. The food to be digested is variously treated to provide a large surface area on which hydrolytic enzymes can act. Different enzymes work best at specific pH's which are found at various regions in the gut. The general pattern of the digestive process is for macromolecules to be broken down into their respective subunits. Carbohydrase, protease and lipase enzymes digest carbohydrates, proteins and fats respectively.

Table 5.1

Site of production	General secretion	Enzyme	Site of action	Substrate	Product	pH
Salivary glands	Saliva	Salivary amylase	Buccal cavity	Starch	Maltose	7.00
		(Mucin also secreted which aids bolus formation)				
Stomach mucosa	Gastric juice	Pepsin(ogen) (pro)Rennin HCl (not an enzyme)	Stomach	Protein Caseinogen Pepsinogen Nucleoprotein	Polypeptides Casein Pepsin Nucleic acid and protein	2.00
		(Mucus also secreted to lubricate food and prevent self-digestion)				
Pancreas	Pancreatic juice	Amylase Trypsin(ogen) Chymotrypsin(ogen) Carboxypeptidase Lipase	Small intestine	Amylose Protein Protein Peptides Fats	Maltose Peptides Amino acids Amino acids Fatty acids + glycerol	7.00
		Nuclease		Nucleic acid	Nucleotides	
Liver	Bile	Bile salts (not enzymes)	Small intestine	Fats (and neutralization of acids)	Fat droplets	
Small intestine	Intestinal juice	Amylase Maltase Lactase ⎫ disaccharases Sucrase ⎭	Small intestine	Amylose Maltose Lactose Sucrose	Maltose Glucose Glucose + galactose Glucose + fructose	8.50
		Nucleotidase Erepsin		Nucleotides Peptides and dipeptides	Nucleosides Amino acids	
		Enterokinase		Trypsinogen	Trypsin	

Table 5.2

Stimulus for secretion	Secretion	Target organ	Effect
Sight, smell, anticipation of food	Saliva	Salivary glands	Conditioned reflex stimulates saliva flow
Food in buccal cavity	Saliva	Salivary glands	Cranial reflex stimulates saliva flow
Food in buccal cavity	Gastric juice	Stomach	Nerve impulses stimulate secretion of gastric juice
⎰ Distension of stomach by food ⎱ Presence of food in stomach	Gastric juice Gastrin	Stomach Stomach	Secretion of gastric juice rich in hydrochloric acid
Presence of fatty acids in food	Enterogastrone	Stomach	Slows peristalsis; inhibits hydrochloric acid production
Food in duodenum	Pancreozymin	Pancreas	Stimulates secretion of pancreatic enzymes
Fatty food in duodenum	Cholecystokinin	Gall bladder	Contraction of gall bladder to release bile
Acidic food in duodenum	Secretin	Pancreas	Stimulates hydrogencarbonate flow in pancreatic juice
		Liver	Bile production

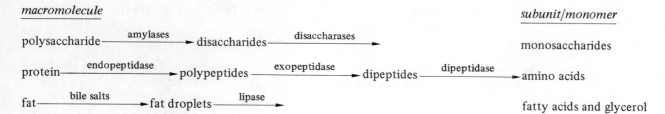

| macromolecule | | | | subunit/monomer |

polysaccharide ——amylases——→ disaccharides ——disaccharases——→   monosaccharides

protein ——endopeptidase——→ polypeptides ——exopeptidase——→ dipeptides ——dipeptidase——→ amino acids

fat ——bile salts——→ fat droplets ——lipase——→   fatty acids and glycerol

The many differentiated regions of the gut all possess the same general structure. This consists of:

1. **Mucosa**: glandular inner layer of the gut secreting mucus and a variety of enzymes. Mucus facilitates easy food movement and prevents self digestion by the enzymes.
2. **Submucosa**: connective tissue containing nerves, blood and lymph vessels.
3. **Muscularis externa**: inner circular and outer longitudinal smooth muscle responsible for **peristalsis**. Circular muscle is specialized into sphincters at various points in the gut. Auerbach's nerve plexus, situated between the two muscle layers, controls peristalsis. Meissner's nerve plexus, located between the circular muscle and submucosa, controls glandular secretion.
4. **Serosa**: outer loose fibrous tissue covered by peritoneum. Peritoneum also forms mesenteries which hold and suspend the gut from the dorsal body wall.

(a) Summary of Chemical Digestion in Man

See Tables 5.1 and 5.2.

(b) Absorption

The **soluble** end-products of digestion are **absorbed** into the body via the **ileum**. This region of the gut possesses numerous thin-walled **villi** which increase its absorptive surface. The cells on the surface of the villi possess **microvilli** which further increase the absorptive surface. Each villus contains a lymph vessel called a **lacteal** and an extensive capillary network. The capillaries transport absorbed food away from the gut and help maintain a steep diffusion gradient. Monosaccharides, amino acids and dipeptides are quickly absorbed into the blood by diffusion or active transport. Fatty acids and glycerol enter the epithelial cells of the villi, are resynthesized into fat, and then passed into lacteals. Protein is added to the fat forming **lipoprotein chylomicrons**. These finally enter the venous bloodstream via the thoracic lymph duct.

Large quantities of symbiotic bacteria reside in the **large intestine** of man, feeding on the remaining undigested food. They synthesize amino acids and vitamin K which are absorbed into the blood. Water is also reabsorbed leaving semi-solid faeces which pass to the rectum. From here they are periodically expelled (**egested**) from the body by peristaltic action.

Once absorbed, food undergoes assimilation as follows:

1. **Monosaccharides**: pass to cells via blood and are used as respiratory substrate; surplus glucose is stored in the liver and muscles as glycogen.
2. **Amino acids**: used in protein synthesis and production of new protoplasm; surplus amino acids are deaminated in the liver. Their NH_2 groups are converted to urea while the remainder of the molecule is converted to glycogen and stored; transamination may occur in the liver.
3. **Fatty acids and glycerol**; stored as fat, a major energy reserve of the body; stored in subcutaneous adipose tissue under the skin, providing insulation, and

around vital organs which are thus protected from buffeting; form a component of cell and nuclear membranes.

Ingestion of food is regulated by the hunger and satiety centres in the hypothalamus. If the level is high the satiety centre is stimulated to inhibit the body from ingesting any further food. When the level is low the hunger centre is stimulated into action.

Carnivores, herbivores and omnivores possess **heterodont dentition**. The general structure of all the teeth is the same but the number, size and shape of them differ according to diet.

Table 5.3

	Man (omnivore)	*Cat (carnivore)*	*Sheep (herbivore)*
Incisors	In upper and lower jaw; flat and chisel-like for biting and cutting	Tear flesh from bone	Upper incisors absent. Replaced by horny pad against which lower teeth bite.
Canines	Pointed; poorly developed.	Highly developed; for piercing and killing prey; fang-like	Upper canines absent. Diastema present—space to keep chewed food from unchewed food.
Premolars	2 cusps on surface; grind and crush food	3rd upper pm, 1st lower m are carnassial teeth; shear flesh. Others flattened with sharp edges for cutting flesh and cracking bones.	Broad grinding surfaces to deal with tough vegetable matter.
Molars	4 to 5 cusps on surface; large surface area for grinding		
Dental formula	$2\left[i\frac{2}{2}\ c\frac{1}{1}\ pm\frac{2}{2}\ m\frac{3}{3}\right]$	$2\left[i\frac{3}{3}\ c\frac{1}{1}\ pm\frac{3}{2}\ m\frac{1}{1}\right]$	$2\left[i\frac{0}{3}\ c\frac{0}{1}\ pm\frac{3}{2}\ m\frac{3}{3}\right]$
Jaw movement	Sideways movement. Up and down movement.	Up and down movement only.	Backward and sideways movement.

In the case of a ruminant (sheep) **symbiotic bacteria** in the **rumen** ferment the ingested food into acetic, propionic and butyric acids which the sheep then uses as energy substrates. The sheep regurgitates and rechews partially digested food from the rumen. It is then reswallowed and passed through a series of compartments to the **abomasum**. This corresponds to the stomach in man. From here digestion is similar to that in man.

5.3 Other Feeding Mechanisms

(a) Microphagous Feeders

These ingest small food particles. Three methods are commonly adopted, as follows.

(i) *Pseudopodial*, e.g. *Amoeba*

Amoeba ingests its food by **phagocytosis**. Pseudopodia surround it and form a **food vacuole**. Intracellular enzymic digestion of the food occurs in the vacuole and the soluble end-products pass into the surrounding cytoplasm by **pinocytosis**.

(ii) *Ciliary*, e.g. *Paramecium*

Tracts of **cilia** in the **oral groove** waft bacteria towards the **cytostome**. Small particles pass into a food vacuole. Digestion occurs in a way similar to that of *Amoeba*. In both cases undigested material is eliminated by **exocytosis**.

(iii) *Filter Feeding*, e.g. Water Flea (*Daphnia pulex*)

Several limbs covered with stiff bristles called **setae**, and protected by a **carapace**, move forward and draw in water containing food particles. The setae filter out the particles. When the limbs move backwards the food is drawn towards the mouth by setae at the base of each limb. Ultimately the food is swallowed after it has been mixed with mucus.

(b) Macrophagous Feeders

These take in large particles. They include the following.

(i) *Tentacular Feeders*, e.g. *Hydra*

As the prey brush against the **cnidocils** of **nematoblasts** located in the tentacles, they cause the discharge of the contents of **nematocysts**. These paralyse, wind round and frequently kill the prey. The tentacle passes the food into the mouth of the *Hydra*. Extracellular digestion occurs in the **enteron**, and the soluble food is then passed into the cells lining the enteron by phagocytosis where further intracellular digestion occurs.

(ii) *Scraping and Boring*, e.g. Snail (*Helix aspersa*)

Vegetation is grasped by the lips, and a toothed **radula** tears off material, breaking up the tough cellulose cell walls, and pressing it against a jaw plate. Food fragments are pushed to the pharynx and swallowed. The cell contents are then digested further along the gut.

(iii) *Biting and Chewing*, e.g. Grasshopper (*Chorthippus*)

Strong, ridged **mandibles** tear up and crush the plant food. The paired **maxillae** and **labium** manoeuvre the food towards the mouth where it is swallowed.

(iv) *Seizing and Swallowing*, e.g. Dogfish (*Scyliorhinus caniculus*)

The ventral wide mouth possesses backwardly pointing **dermal denticles** which prevent the prey from escaping once it has been caught. A tough muscular **tongue** assists the swallowing of whole prey. Folds in the **oesophagus** extend around the swallowed food and prevent water from entering the gut. A **spiral valve**, possessing many infoldings, provides a large surface area for absorption.

(c) Detritus Feeders e.g. Earthworm (*Lumbricus terrestris*)

These eat particles of organic matter. Fragments of vegetation or soil particles are drawn into the **buccal cavity** by the muscular action of the **pharynx**. They are swallowed by peristalsis. The **gizzard** contains small stones which help **masticate** the food. This is then passed into the **straight intestine** where digestion and absorption occur.

(d) Fluid Feeders

These ingest food in liquid form:

    (i)   *Sucking* e.g. Housefly (*Musca domestica*)

Extending from the mouth is a tube called the **proboscis**. It possesses two **labellae** at its distal end each of which contains many **pseudotracheae**. When feeding, the proboscis presses the labellae over the food. Saliva is poured onto solid food which is digested extracellularly. The liquid food passes into the pseudotracheae by capillary action and is then moved up into the gut by the muscular activity of the pharynx.

    (ii)   *Piercing and Sucking*, e.g. Female Mosquito (*Anopheles* sp.)

Stylets extending from the **proboscis sheath** pierce the skin of a mammal and enter a blood capillary. The blood is pumped into the mosquito along a food channel constructed by the **hypopharynx** and **labrum**. An anticoagulant is secreted into the blood to prevent it clotting as it passes along the food channel.

5.3   Saprophytic Nutrition

Saprotrophs are **decomposers** and liberate energy for their own use by breaking down complex organic matter from the dead bodies of other organisms. At the same time this process releases vital chemical elements into the soil which are absorbed by autotrophs. Thus saprotrophs aid the recycling of materials from dead organisms to living ones. Fungal and bacterial saprotrophs are referred to as saprophytes, while animal saprotrophs are called saprozoites.

Mucor hiemalis is a saprophyte and has thin, branched **hyphae**, providing a large absorptive surface. These penetrate dead, decaying matter and secrete enzymes into it. The food is digested extracellularly and is subsequently absorbed and transported to other parts of the fungal **mycelium**.

5.4   Symbiosis

There are two forms of symbiosis, mutualism and commensalism.

(a) Mutualism

Here the relationship is beneficial to *both* organisms. For example, the cellulose-digesting ciliates present in the gut of herbivore ruminants. The ciliates feed on the cellulose. They convert it to simple compounds which the ruminant is then able to assimilate.

(b) Commensalism

Here *one* of the two organisms derives benefit from the other, which itself is not harmed in any way. For example, colonial hydrozoans attach to a whelk shell housing a hermit crab. The hydrozoans feed on the scraps left over after the crab has eaten.

5.5 Parasitism

Ectoparasites such as ticks and fleas live on the outer surface of a host. **Endo-parasites** such as *Taenia* live within the host. **Obligate parasites** must live parasitically all of their lives while **facultative parasites** live initially as parasites, but once their host is dead, continue to feed from it saprophytically. Mistletoe is a partial parasite. It can photosynthesise but at the same time withdraws micro-nutrients from its host via **haustoria**.

Parasites display a wide variety of modifications which adapt them to their highly specialized mode of life. These include structural, physiological and re-productive adaptations.

5.6 Worked Examples

5.1

(a) *Give three classes of organic compound that are essential constituents in the diet of a mammal.*

(i) carbohydrates

(ii) proteins

(iii) fats

[3]

(b) *In a named mammal give the functions of the following regions of the gut:*

Name of mammal Man

(i) *buccal cavity* Food is chewed to increase surface area of food particles. The food is moistened with saliva which contains the enzyme, salivary amylase. This begins the digestion of starch into maltose.

[3]

(ii) *duodenum* Has a large surface area for the secretion of enzymes, including maltase, sucrase and peptidase. Mucus is secreted which aids the movement of food by peristalsis.

[3]

(iii) *colon (large intestine)* Water is absorbed and some vitamins, especially vitamin K. Faeces are formed and stored in the colon.

[3]

(c) Name two energy storage compounds in the mammalian body and state a site where each may be found.

	Energy storage compound	Site in mammalian body
(i)	glycogen	liver
(ii)	fat	beneath skin

[4]

(d) Give the energy storage substances for:

	Name	Storage substance	
(i) a named tap root	sugar beet	sucrose	[1]
(ii) a named seed	maize	starch	[1]

(e) Give two important organic compounds which contain phosphorus and occur in the living cell. For each indicate the function in the cell.

	Name of compound	Function	
(i)	DNA	structure of chromosomes	[2]
(ii)	ATP	energy store	[2]

[OLE]

5.2

You are advised to read the whole of this question before you begin to answer it.

(a) The table below gives three different types of food substance used by a mammal. Indicate the chemical nature of each substance.

Food substance	Chemical nature
(i) Starch	$(C_6H_{10}O_5)_n$
(ii) Lipid	RCOOH, where R=H, CH$_3$, C$_2$H$_5$ etc.
(iii) Protein	chains of amino acids NH$_2$CRHCOOH (R as above)

[6]

(b) In the table below, name an enzyme that initiates the digestive breakdown of each of the food substances listed in part (a) and indicate the optimum pH at which each enzyme works best in the alimentary canal.

	Enzyme	Optimum pH
(i)	amylase	7.0
(ii)	lipase	7.5
(iii)	pepsin	1.5

[6]

(c) In the table below, place the food substances listed in part (a) in order of their calorific value and name a food source in which the particular substance predominates.

Order of calorific value		Food source
(i) highest	lipid	butter
(ii)	carbohydrate	potato
(iii) lowest	protein	meat

[6]

(d) In addition to the food substances given in (a) specify three other essential requirements in the diet of a mammal.

(i) ... water ...

(ii) ... mineral salts ...

(iii) ... vitamins ... [3]

[L]

5.3

The following passage refers to digestion in the mammalian duodenum. Read the passage carefully and then answer the questions that follow.

Peristaltic contractions of the stomach keep the chyme moving towards the duodenum, the first loop of the small intestine. The passage of food into the duodenum is controlled by a ring of muscle situated immediately between the end of the stomach and the duodenum.

The duodenum is the main seat of digestion in the gut. The agents of digestion come from three sources: the liver, pancreas and wall of the intestine. The liver produces bile, a mixture of substances not all of which are concerned in digestion. The digestive components are the bile salts which emulsify fats. It must be stressed that bile salts are not enzymes. Bile is also rich in sodium bicarbonate, which neutralises acid from the stomach. The pH of the small intestine is therefore distinctly alkaline, which favours the action of the various enzymes. Some of these enzymes are constituents of the pancreatic juice which flows into the duodenum from the pancreas via the pancreatic duct. The main pancreatic enzymes are pancreatic amylase, trypsin and pancreatic lipase.

(a) *What are* peristaltic contractions *and what causes them?*

muscular contractions of the wall of the alimentary canal caused by alternate antagonistic contractions of the circular and longitudinal muscles

(b) *Explain the terms* chyme *and* emulsify.

chyme The fluid contents of the stomach. It is composed of small particles of food, water, pepsin, HCl and mucus.

emulsify The process by which small droplets of one liquid are dispersed in another to form a colloidal dispersion.

65

(c) *Name a bile salt.*

sodium glycocholate

(d) *Why are bile salts not classed as enzymes?*

They are not made of proteins nor do they increase the rate of a chemical reaction by lowering the activation energy. They do not have an active site.

(e) *What cells in the stomach produce acid?*

oxyntic cells

(f) *What is the approximate pH of the gastric juice?*

1.5

(g) *What does pancreatic amylase do?*

converts amylose to maltose

[10]

[UCLES]

5.4

(a) *Different stimuli result in secretion of different digestive juices in the mammalian gut. Describe one stimulus for each of*

(i) *saliva* cranial reflex initiated by the presence of food on taste buds

(ii) *gastric juice* distension of stomach walls by the presence of food

(iii) *pancreatic juice* pancreozymin produced by duodenum releases juice

(b) *Describe the role in mammalian digestion of*

(i) *bile salts* Reduce the surface tension of fat globules and cause their emulsification. This aids in the digestion of fat by lipase.

(ii) *enterokinase* Converts inactive trypsinogen into active trypsin which then catalyses proteins into peptides.

[5]

[AEB]

5.5

Samples of 10 cm³ of the stomach contents of a normal person were removed when a meal was given and at half-hour intervals thereafter. The graph (Fig. 5.1) shows the volumes of a 0.1 mol dm⁻³ solution of sodium hydroxide needed to neutralize the acid in the samples.

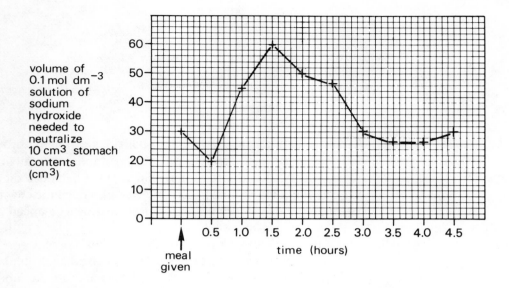

Fig. 5.1

(a) What was the sodium hydroxide equivalent of the basal acid level of the stomach contents? [2]

(b) Why does the acid level fall immediately after eating a meal? [1]

(c) Describe and explain the pattern of acid level of the stomach contents between 1.5 and 4.5 hours after the meal. [3]

(d) Describe the various mechanisms responsible for the increase in acid secretion shortly after the meal. [7]

(e) Name two substances other than acid which are commonly found in the gastric juice of an adult person. [2]

[AEB]

(a) 3 cm³ of 0.1 mol dm⁻³ sodium hydroxide per 1 cm³ of stomach contents.

(b) The addition of food and saliva to the stomach may have neutralized some of the stomach acidity.

(c) The level of acidity in the stomach falls steadily over the first hour from 1.5 hours to 2.5 hours. There is then a much faster fall over the next half-hour and then a steady fall off over the next half-hour to the lowest pH value. This value is then maintained for a further half-hour before rising again slightly. It is essential that there is an adequate amount of acid in the stomach while food is present. This provides the optimum pH of 1.5 which the enzyme pepsin requires in order to break down proteins into polypeptides. The graph shows that the rate of secretion rises to a maximum 1.5 hours after eating a meal but then steadily falls off as food begins to leave the stomach.

(d) There are three mechanisms involved in the secretion of gastric juice. The first is a reflex mechanism initiated by swallowing food. Impulses pass via the vagus nerve to the stomach which secretes gastric juice as food enters the stomach. Secondly, stretch receptors in the stomach wall respond to distension by setting up nerve impulses which lead to the secretion of more gastric juice. Finally, the physical presence of food in the stomach stimulates the gastric mucosa to release a hormone called gastrin which stimulates further production of gastric juice.

(e) Pepsinogen and mucus.

5.6

(a) *Define and explain the importance of a balanced diet.*   **[4]**
(b) *With reference to a* **named** *carnivorous mammal describe how (i) protein is digested and absorbed into the blood stream and (ii) the products are assimilated (used by the body in various ways).*
                    [8, 6]
                    [UCLES]

(a) A diet which contains adequate quantities of energy-providing foods (carbo-hydrates and fats), growth-promoting foods (proteins) and mineral salts, water, roughage and vitamins is said to be a balanced diet. Such a diet would be both adequate and balanced and would reduce the chances of a nutritionally based disease developing. The requirements of a balanced diet change with age and physiological state.

(b) (i) In the lion, protein undergoes both mechanical and chemical digestion. Meat is torn from carcasses using powerful teeth and jaws. The carnassial tooth helps in this action. The surface area of the food is increased by the combined action of incisors, premolars and molars during mastication. Food is then swallowed and passed down the oesophagus by peristalsis. Further mechanical digestion occurs in the stomach as a result of the churning action of the gastric oblique, longitudinal and circular muscles.

In the stomach the inactive enzyme pepsinogen is converted to its active form pepsin by the presence of hydrochloric acid. Pepsin is an endopeptidase and breaks protein chains into polypeptides and peptones.

In the duodenum the pH of the food becomes slightly alkaline due to the release of alkaline fluids from the Brunner's glands. Pancreatic juice, containing trypsinogen, chymotrypsinogen and carboxypeptidases, mixes with the food in the duodenum before passing to the ileum where enterokinase converts inactive trypsinogen into active trypsin. Trypsin then converts inactive chymotrypsinogen into active chymotrypsin. Both these active enzymes convert proteins to amino acids. Erepsin, produced by the ileum, is a mixture of aminopeptidases and dipeptidases and converts peptides to amino acids.

All amino acids are absorbed into the large surface area of the intestinal villi by diffusion or active transport. The amino acids then pass by the hepatic portal vein to the liver.

(ii) According to homeostatic demand there is always a supply of all major amino acids circulating in the blood plasma. These amino acids are used, as required, by cells in the synthesis of proteins to make new protoplasm, repair damaged cells and tissues e.g. muscle, and for the formation of enzymes and hormones. These synthetic activities involve the processes of transcription and translation and require much metabolic energy for the condensation reactions of protein synthesis.

Surplus amino acids cannot be stored and they are deaminated in the liver. The amino groups (NH_2) are removed and converted into urea whereas the remainder of the molecule is converted to glycogen and stored or used up, after conversion to hexose sugar, in respiration.

5.7

Survey the various methods used by invertebrate animals to obtain food.   **[20]**
                    [L]

Invertebrate animals exploit all methods of obtaining food. These include the following methods:

1. Microphagous feeders:
   (a) pseudopodial, e.g. *Amoeba*
   (b) ciliary, e.g. *Paramecium*
   (c) filter feeding, e.g. *Daphnia*
2. Macrophagous feeders:
   (a) tentacular, e.g. *Hydra*
   (b) scraping and boring, e.g. *Helix*
   (c) biting and chewing, e.g. *Locust*
   (d) seizing and swallowing, e.g. *Scyliorhinus*
3. Detritus feeders, e.g. *Lumbricus*
4. Fluid feeders:
   (a) sucking, e.g. *Musca*
   (b) piercing and sucking, e.g. *Anopheles*

For each of these methods a good answer should describe, for named examples, the type of food ingested, the methods of ingestion and any specialized feeding structures. This information is given at the beginning of the chapter. In addition, you should make reference to special methods of feeding such as mutualism, commensalism and parasitism.

6 Respiration and Gaseous Exchange

Respiration   The process by which energy is released by the oxidation of organic molecules in living organisms.

Metabolism   Metabolism is the term which describes the chemical reactions occurring in living organisms. Anabolic reactions result in the synthesis of complex molecules from simpler molecules and energy is required for these reactions. Catabolic reactions result in the breakdown of complex molecules to simpler molecules and energy is usually released in these reactions.

Diffusion   The movement of molecules or ions down a concentration gradient from a region of their high concentration to a region of their low concentration.

Energy   The capacity to do work.

Respiratory quotient   The ratio of the volume of carbon dioxide given off to the volume of oxygen used up over the same time period during respiration.

6.1   Energy

Energy can be neither created nor destroyed. It may occur in a variety of forms, e.g. heat, light, sound, electrical and chemical, and these forms are interconvertible. Energy is stored within cells as chemical energy. This is also the most appropriate form in which it can be transferred efficiently from cell to cell and released in regulated amounts as, and when, required.

All living organisms require a continual supply of energy in order to carry out vital processes. These include:

1. chemical synthesis of substances,
2. growth and division of cells,
3. active transport of substances into and out of cells,
4. electrical transmission of nerve impulses,
5. mechanical contraction of muscle (movement).

Energy in living organisms is derived from the sun. Green plants transduce light energy into chemical energy during photosynthesis. This is an **anabolic** reaction which occurs in **producer** organisms. Other organisms, including animals, are **consumers** and obtain their energy by eating producers. Energy is made available for the functions listed above (1 to 5) during respiration which is a **catabolic** reaction.

6.2   Respiration

Chemical energy is transferred from energy-rich compounds called **respiratory substrates** such as carbohydrates to carrier molecules such as **adenosine triphosphate** (ATP). When the chemical bond between the two terminal phosphate groups of ATP is later broken during hydrolysis reactions, a quantity of energy is released which is sufficient to supply the needs of energy-consuming processes in the cell. Hydrolysis of ATP produces **adenosine diphosphate** (ADP). ADP can be converted back to ATP using 'stored' energy from another 'energy-rich' molecule such as **creatine phosphate (CP)** or by the metabolic processes of respiration.

6.3   Aerobic Respiration

Respiration which requires oxygen is called **aerobic** and that which occurs without the need for oxygen is called **anaerobic**.

Carbohydrates, fats and proteins can all be used as respiratory substrates. Glucose is the major respiratory substrate and during aerobic respiration it undergoes a series of enzyme-controlled oxidation reactions which can be considered in three distinct metabolic phases:

(a) Glycolysis

The reactions of glycolysis are common to both aerobic and anaerobic respiration. Glucose, a hexose (6C) sugar is converted to a **hexose phosphate** molecule using phosphate and energy derived from ATP. The hexose phosphate is split into two **triose phosphate** (3C) molecules which are oxidized to form two molecules of **pyruvic acid** (3C). The oxidation reactions involve the removal of hydrogen atoms and are catalysed by **dehydrogenase** enzymes. A small amount (2 molecules net) of ATP is formed and the hydrogen ions are taken up by the hydrogen acceptor molecule, **nicotinamide adenine dinucleotide (NAD)** to form $NADH_2$.

(b) Tricarboxylic Acid (TCA) Cycle

Each molecule of pyruvic acid has one carbon atom removed by a process called **oxidative decarboxylation**. The carbon dioxide formed is released and the acetyl group (2C) remaining combines with coenzyme A to form **acetyl coenzyme A**. The acetyl coenzyme A then enters a cyclic biochemical pathway called the tricarboxylic acid cycle (or Krebs' cycle) by combining with a 4C compound to form a 6C compound called **citric acid**. At various points in this cycle oxidative decarboxylase enzymes remove carbon atoms. The removed hydrogen atoms are taken up by hydrogen acceptor molecules which channel the hydrogen atoms into a further biochemical sequence called the respiratory chain.

(c) Respiratory Chain (Electron Transport)

Pairs of hydrogen atoms removed from respiratory intermediates by dehydrogenation reactions are split into electrons and protons. The electrons pass along a chain of at least five intermediate molecules including a group of iron-containing carrier molecules called **cytochromes**. Each cytochrome can exist in either an oxidized state (Fe^{3+}) or reduced state (Fe^{2+}) depending on whether or not it is carrying hydrogen ions. At three points of transfer in this respiratory chain sufficient energy is released, in the presence of **phosphorylase** enzymes, to synthesize molecules of ATP from ADP and inorganic phosphate P_i. At the end of this chain an iron- and copper-containing carrier molecule, **cytochrome oxidase**, recombines the electrons and protons and enables the hydrogen atoms to reduce molecular oxygen to water. The overall reaction of the respiratory chain is called **oxidative phosphorylation**.

6.4   Anaerobic Respiration

In the absence of oxygen to accept the hydrogen atom released during glycolysis, pyruvic acid becomes the hydrogen acceptor molecule. Depending upon the metabolic pathways of the organisms, the end-products can be either **ethanol** and **carbon dioxide** as in yeasts (alcoholic fermentation) or **lactic acid** (lactic acid fermentation) as in some bacteria or animal tissues temporarily deprived of oxygen. Both these end-products are toxic if allowed to accumulate. Only 2 molecules of ATP can be directly synthesized from the breakdown of each molecule of glucose during anaerobic respiration as opposed to 38 molecules of ATP during aerobic respiration.

In the subsequent presence of oxygen in animals the **oxygen debt** incurred during lactic acid fermentation is cancelled as lactic acid is converted to pyruvic acid in the liver and finally oxidized in the TCA cycle with the release of more ATP molecules.

6.5   Mitochondria

These rod-shaped organelles approximately $10\,\mu$m long and $1\,\mu$m wide are composed of outer and inner membranes. The inner membrane is folded to form **cristae**. These provide a large surface for the cytochromes and enzymes associated with the respiratory chain. The inner fluid-filled **matrix** of the mitochondria contains the enzymes of the TCA cycle. ATP is synthesized within mitochondria and the number of mitochondria and number of cristae in a cell gives an indication of its metabolic activity. Muscle and liver cells contain large numbers of mitochondria to supply their increased energy requirements.

6.6   Respiratory Quotient (RQ)

The respiratory quotient is 1 for carbohydrates, is 0.7 for fats and lipids and is variable for protein, being generally below 1 and greater than 0.7. The value of RQ indicates the type of substrate being used for respiration. An RQ greater than 1 indicates that both anaerobic and aerobic respiration are taking place. In plants, low RQ values indicate that respiration and photosynthesis occur simultaneously.

6.7   Gaseous Exchange

Aerobic animals remove oxygen from their environment and return carbon dioxide. This exchange occurs by **diffusion**. The region of the organism through which these gases are exchanged must have a *large surface area*, be *thin* and be *permeable* to respiratory gases. In small organisms such as bacteria and protozoa where the surface-area-to-volume ratio is large the outer cell **membrane** serves as the **gaseous exchange surface**. The gases have a short distance to travel from environment to mitochondria and the metabolic demands of the organism are low.

Simple multicellular organisms such as coelenterates and platyhelminthes continue to rely on gaseous exchange through their outer surfaces. In both cases their increases in volume are accompanied by a dorso-ventral flattening of the body so as to maintain high surface-area-to-volume ratios. All more advanced animals require specialized gaseous exchange structures and mechanisms to compensate for the increases in size and metabolic demand. In nearly all these groups of animals, apart from insects, the blood vascular system acts as a transport medium conveying respiratory gases from the exchange surface to respiring tissues. The presence of **respiratory pigments** such as **haemoglobin** increases the efficiency of the transport system.

Water contains less oxygen per unit volume than air, therefore an aquatic animal has to expend a great deal of energy in passing a large volume of water over its exchange surface. Terrestrial animals do not need to move such a large volume of air. Terrestrial animals, however, have to contend with problems of water loss to the atmosphere. Any surface which is permeable to gases is also permeable to water, hence the need for gaseous exchange surfaces to be situated inside the body where they are protected from dehydration.

6.8   Gaseous Exchange Surfaces

Annelids have a fairly well-developed blood vascular system and blood containing respiratory pigment. Gaseous exchange occurs through the **epidermis** (*Lumbricus*), **external gills** (*Arenicola*) or **parapodia** (*Nereis*).

Insects have a **tracheal system** composed of paired external openings called **spiracles** controlled by valves and which lead to a series of tubes called tracheae. The tracheae end in smaller tubes called **tracheoles** which supply the most active metabolic tissues. The tracheae are lined with rings of chitin for support but these are absent from tracheoles. The terminal regions of the tracheoles contain fluid through which the respiratory gases must pass to the tissues. In some larger insects **ventilation movements** of the abdomen speed the flow of air through the tracheal system.

Fish have **gills** situated in **gill slits**. In dogfish each slit remains separate but in bony fish these are covered by a moveable cover called the **operculum**. The surface area of the gills is increased by folds called **lamellae** which have **secondary lamellae** or **gill plates** projecting from them. Gaseous exchange occurs between the blood in the lamellae and the water. The rate of water flow over the lamellae is increased by ventilation movements involving the **buccal cavity** and either the **pharyngeal cavity** (dogfish) or the **opercular cavity** (bony fish). In bony fish a **countercurrent system** increases the efficiency of oxygen uptake.

Mammals have a pair of lungs situated in the **thoracic cavity**. *Volume* changes brought about by muscular contractions of the **diaphragm** and **intercostal muscles** produce *pressure* changes in the **thorax** which result in ventilation of the lungs. Air, drawn in through the **mouth** and **nostrils**, passes through the **pharynx**, **larynx**, **trachea**, **bronchi** and **bronchioles** before the oxygen diffuses through the **alveoli** into the blood capillaries. Carbon dioxide diffuses out of the body in the reverse direction. **Inspiration** of air is an active process controlled by the **inspiratory centre** in the **medulla**. It is stimulated into activity by **chemoreceptors** which detect the **carbon dioxide concentration** of the blood. **Expiration** is largely a passive process brought about by the elastic recoil of lung tissue and respiratory muscles.

6.9   Worked Examples

6.1

(a) *Give an equation that summarizes (i) aerobic respiration and (ii) anaerobic respiration.*

(i) *aerobic respiration*

$$C_6H_{12}O_6 + 6O_2 \longrightarrow 6CO_2 + 6H_2O + 38ATP$$

(ii) *anaerobic respiration*

$$C_6H_{12}O_6 \longrightarrow 2CH_3CH_2OH + 2CO_2 + 2ATP$$

[2]

(b) *What might induce anaerobic respiration in (i) parenchyma cells in a plant root and (ii) mammalian striped muscle fibres?*

(i) *parenchyma cells in a plant root*

lack of oxygen supply in waterlogged soil

(ii) *mammalian striated (skeletal) muscle fibres*

inadequate supply of oxygen during vigorous activity

[2]

(c) *For each of the following respiratory quotient values in a green plant, state the type of respiratory substrate being used and the conditions in which the process occurs.*

Respiratory quotient	Respiratory substrate	Conditions in which process occurs
1.0	carbohydrate	total darkness
0.7	protein	prolonged darkness - no available carbohydrate left
0.5	oil	germination of lipid-rich seeds

[6]

(d) *Why are high respiratory quotient values obtained from tissues involved with conversion of carbohydrate to fat?*

The conversion of carbohydrate to fat releases carbon dioxide. A lot of CO_2 released cos it is broken down to AcetyCoA compared to amount of O_2 taken in

[3]
[L]

6.2

Read through the following account of cellular respiration and then write on the dotted lines the most appropriate word or words to complete the account.

The initial phase in the breakdown of glucose, a process known as ... glycolysis ...,

takes place anaerobically in the general cytoplasm of the cell and eventually results in the

production of two molecules of ... pyruvic acid ... from each molecule of

glucose. In most organisms this product then enters the second phase of cellular respira-

tion, known as the ... tricarboxylic acid ... cycle, which occurs in specific organelles,

74

the . . . . *mitochondria* . . . ., under aerobic conditions. Hydrogen is removed from the substrate and passes through a sequence of steps in which carrier molecules, or *coenzymes* . . . . , are involved. The intermediate reactions in this sequence do not require molecular oxygen and are catalysed by. . . . . . . *dehydrogenase* . . . . enzymes. The final stage in the hydrogen carrier system is completed in the presence of oxygen and is catalysed by the haem-containing enzyme. . . . . *cytochrome oxidase* . . . . . This enzyme is sensitive to respiratory poisons such as. . . . . *cyanide* . . . . In the respiratory process, energy is released and is used to synthesise energy rich molecules of . . . . . . *ATP* . . . . from . . . . *ADP* . . . ., thereby storing energy for future use. The output of these molecules in the aerobic phase, a process known as . . . . *phosphorylation* . . . . ., is greater than their output in the anaerobic phase. In evolutionary terms, however, the latter phase is the older established and can provide the sole source of energy in organisms such as bacteria via the process of . . *anaerobiosis* . . .

6.3

Explain the biochemistry of the following processes and state where each process you describe occurs within the cell:
(a) glycolysis;
(b) citric acid (Krebs') cycle;
(c) oxidative phosphorylation.

(a) *Glycolysis*
The biochemical detail required in this answer is stated in section 6.3(a). These chemical reactions occur within the cytoplasm of the cell.
(b) *Citric acid (Krebs') cycle*
This is also known as the tricarboxylic acid (TCA) cycle and details of the cycle are described in section 6.3(b). The chemical reactions of this cycle are catalysed by enzymes situated in the matrix of the mitochondrion.
(c) *Oxidative phosphorylation*
The biochemical details of this process are described in section 6.3(c). The reactions occur on the cristae of the mitochondrion and may involve elementary particles found on the matrix side of the cristae.

6.4

Outline the chemical changes which proteins and fats undergo when they are used as an energy source. Indicate where the products of these changes link up with carbohydrate respiration.
[6]

Proteins are hydrolysed into amino acids and deaminated by oxidation, by deamination, or by transamination. In the former case an α–keto acid is formed. This may be converted into pyruvic acid or into acetyl CoA and undergo respiratory reactions as would either of these compounds had they been derived from carbohydrate sources.

The addition of an amino group from one amino acid to a keto acid can form new amino acids. In the process an α–keto acid is formed which may enter the TCA cycle as say oxoglutaric acid or as oxaloacetic acid.

Fats are hydrolysed into fatty acids and glycerol. Fatty acids undergo β-oxidation to form acetyl CoA which enters the TCA cycle. Glycerol is converted into triose phosphate which is incorporated into the glycolysis pathway.

6.5

In the space below make a diagram to show the structure of a mitochondrion as revealed by the electron microscope.

(a) Label four of its component parts. [2]

(a)

matrix   outer membrane

crista   inner membrane

(b) └────────────────┘
    5 μm

(b) Add a scale to your diagram which shows the approximate size of the organelle. [1]

(c) How would you proceed to isolate mitochondria from the cells of a living tissue such as the brain or liver?

Break the tissue down by cell homogenization and separate out mitochondria by ultracentrifugation. [2]

(d) Having obtained a suitable sample of mitochondria in (c) above, briefly describe how you could demonstrate any one of their functions experimentally.

Mitochondria contain dehydrogenase enzymes of the TCA cycle. Add mitochondria to a buffered sucrose and succinic acid solution in the presence of DCPIP, a hydrogen acceptor, which loses its blue colour as it is reduced. Loss of colour indicates breakdown of succinic acid to fumaric acid.

[4]

[OLE]

6.6

(a) State three features of respiratory surfaces which are common to all vertebrate animals, and briefly explain why each is important.

first feature thin

importance gases can only diffuse quickly over short distances

second feature permeable to gases

importance respiratory surfaces must have pores through which molecules of gas can diffuse

third feature large surface area
....

importance to increase the rate at which gases diffuse
across the respiratory surface
....

[6]

(b) *Explain the possible effects of a decrease in environmental temperature on the rate of gas exchange in*

(i) *a well-illuminated foliage leaf*

Respiration and photosynthesis are enzyme-controlled reactions - enzyme activity is reduced by lower temperatures. The rate of gas exchange in the light, i.e. CO_2 in and O_2 out, will be reduced.

[2]

(ii) *a small mammal*

In order to maintain a constant body temperature the metabolic rate must rise to balance increased heat loss. Thus the respiration rate would increase.

[2]

(c) *Explain the possible effects of a decrease in light intensity on gas exchange in a previously well-illuminated foliage leaf.*

If light intensity was limiting, the rate of photosynthesis would decrease. CO_2 output would increase and O_2 output would decrease and the compensation point might be reached. Stomatal apertures may close.

[4]

[L]

6.7

Figure 6.1 shows a section of mammalian lung. Identify parts 1 to 4, giving a reason in each case.

1. terminal bronchiole - small-diameter passage

2. alveoli - thin-walled epithelia interspersed with capillaries

3. TS bronchiole - wall of cartilage and smooth muscle

4. large artery - thin tunica intima, thick tunica media of smooth muscle cells

[4]

[AEB]

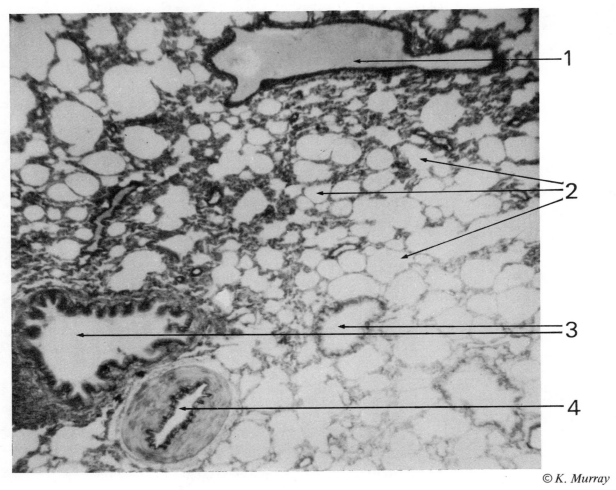

Fig. 6.1

6.8

Compare the respiratory exchange systems of a mammal and a fish describing:

(a) the structures in which gaseous exchange takes place;
(b) the way in which the surrounding medium is brought to the exchange surfaces;
(c) how gases are transported about the body;
(d) the responses to varying respiratory demands.

(a) Gaseous exchange surfaces of mammals and fish have the same basic features: thin, large surface area permeable to respiratory gases, and close to a transport system. Gaseous exchange takes place in alveoli in mammals. These are formed from groups of simple epithelial cells and blood capillaries. Air fills the hollow spaces of the alveoli and gaseous exchange occurs between the alveolar spaces and the blood in the capillaries. The structure of the gills of fish is fundamentally different from alveoli in mammals. Large numbers of gill lamellae project out at right angles from the surface of the gill filaments. Blood in capillaries flows through the lamellae and gaseous exchange occurs by diffusion between the water flowing over the lamellae and the blood in capillaries.

(b) Ventilation movements, produced by muscular activity in both mammal and fish, ensure an almost continuous flow of air in the case of the mammal, and of water in the fish, to the exchange surfaces. Thus steep diffusion gradients are maintained between the environment and the blood capillaries.

In both mammal and fish, increases in volume in certain body cavities, thoracic in mammals and buccal in fish, result in a decrease of pressure within those cavities which leads to a movement of the surrounding medium, air or water, into the organism.

Muscular contractions of the diaphragm and external intercostal muscles cause the volume of the thorax to increase. Air, containing approximately 21 per cent oxygen, is forced into the lungs in a single stage due to the fall in atmospheric pressure in the lungs. In both cartilaginous and bony fishes contraction of hypobranchial muscles lowers the floor of the pharynx and produces volume changes which lead to water being drawn into the pharynx. A second stage of muscular contraction then raises the floor of the pharynx and forces water, containing oxygen, over the gills on its exit from the body. The precise mechanism differs between cartilaginous and bony fish, owing to the presence of discrete gill slits in the former and an operculum in the latter.

(c) There are striking similarities in the methods by which gases are transported about the body in mammal and in bony fish. In both cases respiratory gases are transported from gaseous exchange surfaces to actively respiring tissues by a blood vascular system. Furthermore, the blood of both transport systems contains a respiratory pigment which increases the efficiency of the blood's oxygen-carrying capacity. In both cases the pigment is haemoglobin, although owing to differences in the molecular configuration of mammalian haemoglobin and fish haemoglobin, the former can carry approximately two-and-a-half times the amount of oxygen of the latter.

The methods of transport of carbon dioxide are similar in both animals. Some carbon dioxide is transported as hydrogencarbonate ions, some as carbamino compounds and some in physical solution.

The speed of circulation of blood is greater in mammals than in fish in order to satisfy the higher metabolic demands of homeothermic mammals. Oxygenated and deoxygenated blood are kept separate in the mammal, which has a double circulation as opposed to the single circulation of fish. This difference is also associated with a difference in heart structure between the two organisms.

(d) Varying respiratory demands, i.e. the varying energy requirements of organisms, are influenced by changes in environmental factors, e.g. temperature, or internal factors such as muscular activity, stage of growth and development and breeding. In both fish and mammal, increased demands for energy are detected by the nervous and endocrine systems which increase the supply of oxygen and the removal of carbon dioxide, to and from, respectively, the gaseous exchange surface. These homeostatic mechanisms are controlled by negative feedback. The influence of homeothermy in mammal and poikilothermy in fish are major determinants of the extent of the response shown by the two groups of organisms.

7 Transport

Mass flow The bulk transport of materials from one point to another as a result of a pressure difference between two points.
Osmosis The movement of water or solvent molecules from a region of their high concentration to a region of their low concentration through a differentially permeable membrane.
Osmotic pressure (OP) The pressure a solution generates when enclosed within an osmometer and allowed to come to equilibrium with pure water.
Transpiration The loss of water vapour from the surface of the plant. It may occur via stomata in leaves, lenticels in stems and the cuticle.
Guttation The loss of liquid water from the surface of a plant.
Antibody A molecule synthesized by an animal in response to the presence of foreign substances for which it has a high affinity.
Antigen Foreign material that elicits antibody formation.

Unicellular organisms and individual cells of multicellular organisms possess a large surface-area-to-volume ratio. Efficient exchange of materials between these cells and their environment takes place by diffusion, active transport and endo- and exocytosis. The distances that materials travel within the cell are small enough for diffusion and cytoplasmic streaming to be fast and efficient. However, in the bodies of large multicellular organisms different cells are far apart from each other and/or their external environment and diffusion is inadequate. Thus long-distance mass-flow transport systems have been developed.

7.1 Transport in Plants

The movement of materials through the vascular tissues of plants is called **translocation**. **Xylem** tissue transports water and dissolved mineral salts from the roots to the rest of the plant, while **phloem** translocates dissolved organic and inorganic solutes.

(a) Understanding Osmosis

If a **hypertonic** solution is separated from a **hypotonic** solution by a **differentially permeable membrane**, a net flow of water molecules occurs through the membrane from the hypotonic to the hypertonic solution. When there is no longer a net flow of water both solutions are **isotonic** with each other. The higher the solute concentration the higher its osmotic pressure (OP). However, this OP is a potential value and is more correctly termed osmotic potential (π). By convention, π is given a negative sign.

(b) Water Potential (ψ)

This is the tendency of water to enter or to leave a system. Water moves from a region of higher ψ to one of lower ψ. ψ for pure water at atmospheric pressure is zero. All solutions at this pressure have lower ψ's than pure water and therefore negative values. As a solution becomes stronger its ψ becomes more negative.

(c) Osmosis and Plant Cells

When a cell is placed in a solution of lower ψ than itself, water leaves the cell (**exosmosis**) and the cell contents begin to shrink away from the cell wall (**incipient plasmolysis**). At this point the cell is **flaccid** and no pressure is exerted by the protoplast against the cell wall. If water continues to leave the cell the plasma membrane is pulled away completely from the wall and the cell becomes plasmolysed.

If the cell is placed in a solution of higher ψ than itself, water enters the cell (**endosmosis**). As the protoplast and vacuolar volume increases it exerts an increased pressure on the cell wall and prevents the elastic wall from contracting inwards. This is called **turgor pressure** and is equal and opposite to the **wall pressure** exerted on the protoplast. A cell is in a state of **turgor** when it reaches its maximum turgor pressure. At this point the tendency for water to enter the cell is balanced by the amount leaving it and is zero.

The quantities of water exchanged between cells are insufficient to affect markedly their osmotic pressures though turgor pressures can change significantly.

Water moves through a plant along a gradient of water potential from a region of higher ψ in the soil to the atmosphere, which has a lower ψ. Solar energy causes water to evaporate from leaves and thus maintains the movement of water through the plant.

As water is **transpired** from the leaf mesophyll cells they develop a lower ψ than the leaf xylem sap. Consequently, water from the xylem moves along a gradient of water potential to replace the evaporated water. This places the water columns in the elongated xylem tracheids and vessels of the stem under tension. The polar water molecules stick to each other (**cohesion**) and to the walls of the xylem elements (**adhesion**). Such is their strength of attraction that they remain together when the tension in the vessels pulls the water column upwards by mass flow (**transpiration stream**).

In the root, solutes are actively secreted into the xylem sap by neighbouring cortex and endodermal cells. This, together with the constant removal of water from the root to the stem, produces a lower ψ in the xylem than in the soil and water moves from the soil across the root to the xylem.

Water moves across the root and leaf by three different pathways:

1. Through the **apoplasm** which is a system of cellulose cell walls. Much of the free space in cellulose is occupied by water. The cohesion properties of water aid a continuous flow of water along an osmotic gradient.
2. Through the **symplasm**. This is where the cytoplasm of adjacent cells is linked by plasmodesmata. Water passes along this system down a water potential gradient. It may be further aided by cytoplasmic streaming.
3. Water moves down a water potential gradient from one adjacent cell vacuole to another. The water crosses the apoplasm and symplasm and also moves through the plasma membrane and tonoplasts by osmosis.

Whichever pathway the water takes in the root, it finally reaches the **endodermis**. Here the waterproof **Casparian strip** impedes any further progress. In order for the water to reach the xylem it must pass into the cytoplasm of the endodermis and therefore come under direct cytoplasmic control. The endodermis also prevents backflow of water.

Stomata permit gaseous exchange between the plant and its environment and also control the rate of water loss. A traditional view of how stomata operate is the **sugar–starch hypothesis**. This suggests that in light, guard cells produce sugar which decreases their ψ. This causes endosmosis and increases the turgidity of the guard cells which then cause the stoma to open.

A modification of this idea suggests that when photosynthesis resumes, carbon dioxide is used up and the pH is raised. This stimulates enzymes to convert starch to sugar which ultimately leads to stomata opening.

A more recent theory states that opening of stomata in response to light is achieved by K^+ ions and associated organic anions being actively pumped into the guard cells. This increases the solute concentration and promotes endosmosis. Malate may be the anion component as it is known that in some plants guard-cell starch grains are converted into malate in the presence of light.

Most of the water lost from a plant is transpired via the leaf stomata. The heat energy lost during transpiration helps cool the plant, and the transpiration stream aids both distribution of mineral salts throughout the plant and their uptake from the soil.

External and internal factors affect the rate of transpiration:

(i) *External (Environmental) Factors*

1. **Light**   With few exceptions stomata open in the light and close in darkness. When open the rates of gaseous exchange and transpiration are increased.
2. **Temperature**   As temperature increases the atmospheric humidity decreases. This induces a greater rate of transpiration.
3. **Humidity**   High humidity decreases the diffusion gradient of water between the atmosphere and substomatal space and reduces transpiration rate.
4. **Wind**   This sweeps away the shell of still air round a leaf, thus increasing the water diffusion gradient and therefore the rate of transpiration.
5. **Soil water**   If water becomes increasingly unavailable then less is taken in and transpiration is reduced.

(ii) *Internal Factors*

Transpiration increases as the surface area increases or if the surface-area-to-volume ratio increases. The thinner the cuticle the greater the rate of cuticular transpiration. Transpiration rate also increases the more stomata there are per unit area.

Mineral elements are absorbed as ions from the soil solution via **root hairs**. They enter passively by diffusion or are selectively absorbed by active transport. Once in the root they move towards the xylem via the apoplast and symplast pathways. From the root xylem (the 'source') they pass by mass flow to sites of utilization ('sinks'), e.g. growing regions. After their initial delivery, the ions may be recirculated to other parts of the plant by the phloem.

(d) Translocation of Organic Solutes

One theory suggests that the soluble products of photosynthesis are passed from the leaf ('source') into phloem elements by active transport. Elsewhere in the plants these solutes are actively absorbed into cells ('sinks') and utilized. Therefore a hydrostatic pressure gradient exists between the source and sinks resulting in the passive movement of solutes by mass flow in the direction of the sinks.

The **electro-osmosis theory** agrees that solute movement in the phloem is by **mass flow** but states that it is boosted by electro-osmotic forces operating across the **sieve plates**. The potential difference across a sieve plate is maintained by a protein pump with energy for its operation supplied by companion cells. At a critical potential difference a surge of protons from the sieve tube walls temporarily reverses the potential difference and causes the passage of K^+ through the sieve plate. This electro-osmosis of K^+ ions induces the mass flow of the solution.

7.2   Transport in Animals

Transport systems in animals generally consist of a circulating fluid, the blood, a system of tubes in which the blood is transported, and a contractile mechanism, the heart, which maintains blood flow around the body.

The earthworm possesses a **closed** vascular system. Here the blood is confined to a series of blood vessels and not permitted to mix with the body tissues. Blood is pumped around the system by muscular **longitudinal dorsal** and **ventral vessels** and five pairs of lateral **pseudohearts** in segments 7 to 11. Backflow is prevented by **valves** and **capillary networks** permit exchange of materials between tissues and blood. The blood itself contains **haemoglobin** dissolved in plasma and some phagocytic cells.

Insects have an *open* vascular system. Typically they possess a single dorsal blood vessel which transports blood anteriorly by the action of valves in the 'heart' region of this blood vessel. Circulation through the blood vessel depends upon muscular activity of the body. The internal organs are suspended in a network of blood-filled sinuses which collectively form the **haemocoel**. There are no other blood vessels. The blood itself is colourless, possessing **amoeboid leucocytes** but no haemoglobin.

All vertebrate systems possess a prominent muscular heart which pumps blood around the body. **Arteries** transport blood away from the heart. Those near the heart possess elastic walls to withstand the pressure of the cardiac output. They stretch during ventricular systole and recoil during diastole. This gives rise to the characteristic **pulse**. Further from the heart, arterioles possess walls of smooth muscle. Vasodilation and vasoconstriction of these vessels constantly adjust blood-pressure and the distribution of blood throughout the body. Precapillary sphincters and arterio-venous cross-connections regulate blood-flow through capillaries according to the needs of the tissues or organs concerned.

Capillaries link arterioles to venules. They are small, thin-walled blood vessels and the site of exchange of materials between blood and tissues. Blood moves slowly in them and they present a large surface area.

Veins receive blood at a low non-pulsatile pressure and transport it towards the heart. Their walls are relatively non-muscular, thinner and less elastic than arteries. Pocket valves maintain a one-way flow of blood in them.

The dogfish has a two-chambered heart (one atrium and one ventricle). Deoxygenated blood from the body is pumped by the heart to the gills. Here it is oxygenated before passing around the body and ultimately returning to the heart. As blood only flows through the heart once during one complete circuit of the body this system is called a **single circulation**. Blood-pressure is lowered considerably when blood passes through the gills and consequently blood-flow is slow.

The frog exhibits a partial **double circulation**, possessing a heart with two atria and one ventricle. Blood from the body enters the right atrium and is pumped to the lungs by the common ventricle. It returns to the heart and enters the left atrium before being pumped around the body. A spiral valve in the **conus arteriosus** helps to keep deoxygenated and oxygenated blood separate to some extent.

Mammals have a four-chambered heart and a complete double circulation. Blood must pass through the heart twice during one complete circuit of the body and this helps maintain a high blood-pressure and therefore fast blood-flow.

A **cardiac cycle** consists of one **systole** and one **diastole**. During diastole venous blood enters the **atria**. As they become extended pressure builds up in them. This forces open the **bicuspid** and **tricuspid** valves and some blood flows into the **ventricles**. When diastole ends, the two atria contract (atrial systole), forcing more blood into the ventricles. Next the ventricles contract (ventricular systole), and, with the bi- and tricuspid valves closed, blood is forced into the main arteries. Once in them pocket valves prevent backflow. Systole is then followed by diastole where the cardiac muscle relaxes and the heart refills with blood.

Cardiac muscle is **myogenic** and its rhythmic beat originates in the **sino-atrial node (SAN)** located in the right atrium. Waves of electrical excitation from the SAN pass to both atria, causing them to contract. Excitation then extends to the two ventricles via the **atrio-ventricular node**, **bundle of His** and **Purkinje tissue**. Both ventricles then contract simultaneously. The total volume of blood pumped through the heart per minute is called the **cardiac output**.

Heartbeat activity is modified by the cardiovascular centre located in the medulla oblongata of the brain. Nerves from the autonomic nervous system pass from here to the SAN. Impulses conveyed by the sympathetic system speed up the heart-rate while impulses passing along the vagus nerves of the parasympathetic system slow down heartbeat. Collectively these systems regulate heart-rate homeostatically according to the needs of the body. Non-nervous stimuli also modify heartrate. For example, low pH and high temperatures accelerate it, while high pH and low temperatures decelerate it.

Nervous control of blood pressure is carried out by **baroreceptors** located in the **aorta** and **carotid arteries**. When blood-pressure is low they stimulate the **vasomotor centre (VMC)** of the brain to send impulses to arterioles via sympathetic nerve fibres. This induces **vasoconstriction** which causes increased resistance to blood flow and a corresponding rise in blood-pressure. Conversely, when blood-pressure is high, impulses from the VMC pass along parasympathetic fibres and stimulate **vasodilation** which causes a reduction in blood-pressure.

Chemically, high carbon-dioxide levels stimulate the VMC to vasoconstrict arterioles. The resulting higher blood-pressure transports carbon dioxide more rapidly to the lungs for expulsion and exchange with oxygen. Where tissues suddenly become active they produce more carbon dioxide. This causes vasodilation of local blood vessels, thus increasing their blood supply and allowing more oxygen and glucose to reach them for respiratory purposes. Emotional stress and adrenaline both promote vasoconstriction and increased blood pressure.

(a) Composition of Blood

Blood is composed of 55 per cent fluid plasma and 45 per cent cells. **Plasma** contains water (90 per cent), plasma proteins, the clotting agents **prothrombin** and **fibrinogen**, enzymes and a variety of mineral ions. While these components are maintained at a constant concentration, soluble food and excretory materials, vitamins and hormones occur in varying amounts.

In mammals, **erythrocytes** are small enucleate biconcave discs containing haemoglobin. Their large surface-area-to-volume ratio is efficiently exploited for gaseous exchange. Their pliable membranes enable them to squeeze through capillaries with internal diameters smaller than their own. Erythrocytes function for about three months and are then destroyed in the spleen or liver.

Leucocytes possess a nucleus, are larger than erythrocytes and play an important role in the defence of the body. Two types exist, granulocytes and agranulocytes. **Granulocytes** are further classified as **neutrophils** (phagocytes), **eosinophils** which possess anti-histamine properties, and **basophils** which manufacture heparin and histamine. **Agranulocytes** consist of phagocytic **monocytes**, and **lymphocytes** which promote immune reactions. **Platelets** formed from megakaryocytes in the bone marrow initiate the mechanism of blood clotting.

Tissue fluid is formed when fluid from the plasma is forced out of the blood capillaries under pressure. It bathes the cells and is the medium through which exchange of materials between blood and tissues occurs. Some tissue fluid is reabsorbed into the venous end of the blood capillaries, while the rest enters lymphatic capillaries to become **lymph**. This is moved through larger lymph vessels by contraction of the muscles surrounding them, and backflow is prevented by valves. Lymph is returned to the blood via the right subclavian vein. **Lymph nodes** occur at intervals in the lymphatic system. They contain lymphocytes which produce antibodies and play a role in the body's defence against infection.

(b) Functions of Mammalian Blood

1. The transport of soluble food, excretory materials, other metabolic by-products and hormones in its plasma.
2. The maintenance of a constant blood pH and OP by plasma protein activity.
3. The distribution of heat from deep-seated organs to the rest of the body.
4. The transport of oxygen from the lungs to all parts of the body, and carbon dioxide from the tissues back to the lungs.
5. Defence against disease.

(i) *Oxygen Carriage*

This is the function of the erythrocytes which contain haemoglobin. Where the **partial pressure (pp)** of oxygen is high, as in the lungs, oxygen readily combines with haemoglobin to form unstable **oxyhaemoglobin**. This circulates around the body and in regions of low partial pressures (pp) of oxygen, e.g. tissues and organs, oxygen is released and diffuses into the surrounding cells.

Haemoglobin is completely saturated with oxygen at a point called its **loading tension**. When the percentage saturation of blood is plotted against the pp of oxygen, an S-shaped curve called the **oxygen dissociation curve** is obtained. In regions of increased pp of CO_2 the oxygen dissociation curve is shifted to the right and oxygen released more readily. This is the **Bohr effect**. Over the steep part of the curve, a small decrease in the pp of oxygen in the surrounding atmosphere brings about a large fall in the percentage of oxygen saturation of haemoglobin and the oxygen released becomes available to the tissues.

(ii) *Carbon Dioxide Transport*

Carbon dioxide is carried by the blood in solution (5 per cent); combined with protein as carbamino-haemoglobin compounds (10 to 20 per cent), and as hydrogencarbonate (85 per cent). The latter involves interactions between water, carbon dioxide, haemoglobin, sodium chloride and potassium ions. During the process haemoglobin acts as a buffer molecule accepting H^+ ions. This permits large quantities of carbonic acid to be carried to the lungs without any alteration in blood pH. Hydrogencarbonate ions formed in erythrocytes diffuse into the plasma and form sodium hydrogencarbonate. Loss of hydrogencarbonate ions is balanced by chloride ions diffusing into the erythrocytes. This is called the **chloride shift**.

(c) Defensive Functions

(i) *Clotting*

This process depends on at least twelve clotting factors. Essentially, damaged blood cells and platelets release **thromboplastin**. In the presence of Ca^{2+} ions this catalyses the formation of the enzyme thrombin from its inactive precursor **prothrombin**. **Thrombin** catalyses the conversion of **fibrinogen** to **fibrin** to form the clot.

(ii) *Phagocytosis*

Plasma proteins called **opsonins** become attached to bacteria and in some way help neutrophils recognize them. The bacteria are then engulfed in amoeboid fashion by phagocytes. A **phagosome** is formed and the bacteria digested. Neutrophils can squeeze through the blood capillary walls (**diapedesis**) and move through tissue spaces. Macrophages resident in the liver, spleen and lymph nodes engulf toxic foreign particles and help localize infection.

(iii) *Immune system*

Two systems of immunity exist:

1. *Cell-mediated*: membrane receptors of **T-cells** recognize an antigen and proliferate a clone of T-cells. These then directly combat the antigens.
2. *Humoral*: **B-cells** recognize an antigen and proliferate to form a plasma cell clone. These cells liberate antibodies into the plasma. The antibodies adhere to the bacterial surface and either help speed up phagocytosis or neutralize toxins produced by microorganisms.

(d) Blood Groups

Erythrocyte cell membranes possess **agglutinogens (antigens)** which may react with **antibodies (agglutinins)** contained in the plasma. Two agglutinogens **A** and **B** exist, as do two complementary agglutinins **a** and **b**. If donor erythrocytes are incompatible with a recipient's plasma, agglutination occurs. However there is little effect if the recipient's erythrocytes are incompatible with the donor's plasma. Blood group O individuals are universal donors while AB individuals are universal recipients (see section 16.7).

7.3   Worked Examples

7.1

(a) *Explain the meaning of the following terms:*
(i) osmotic potential (osmotic pressure)
(ii) water potential (diffusion pressure deficit)
(iii) turgor pressure.   **[10]**
[L]

(b) *Describe with full experimental details how you could determine the water potential (diffusion pressure deficit) of plant tissue such as potato tuber or beetroot.*   **[10]**

(a)   (i)   Osmotic potential (π) is the potential pressure which would be exerted between two solutions if they were separated by a differentially permeable membrane. Pure water has an osmotic potential of O. As solute molecules are added to water the osmotic potential decreases, i.e. becomes more negative.

(ii)   Water potential (ψ) for most practical purposes is the potential pressure that exists between the solution within a living plant cell and a surrounding liquid:

$$\text{water potential of cell} = \text{pressure potential of the cell system} + \text{osmotic potential of cell solutions}$$

Water molecules will move from a region of high ψ to a region of low ψ through a differentially permeable membrane.

(iii)   Turgor pressure is the pressure exerted by the protoplast of a plant cell on the cell wall. It is equal and opposite to the wall pressure and prevents the elastic wall from contracting inwards. Turgor pressure is generated by the expanding vacuolar volume produced by the movement of water into the cell in response to the water potentials of the cell and the surrounding liquid.

(b)   Experiments to determine the water potential of a piece of potato tuber.

METHOD
1. Prepare a series of sucrose solutions of known molarity from 0.1 M to 0.6 M at 0.1 M intervals.
2. Place the same volumes of each solution into 6 labelled petri dishes and cover. Set up another petri dish containing the same volume of distilled water, cover and label.
3. Cut 28 potato chips exactly 0.5 cm x 0.5 cm x 5 cm long.
4. Dry all cut surfaces on blotting paper.
5. Add 4 potato chips to each of the labelled petri dishes containing the prepared liquids.
6. Leave the potato chips completely immersed in the liquids for one hour.
7. Remove the potato chips and measure accurately their lengths to the nearest 1 mm.
8. Calculate the mean change in length of the chips in each of the solutions.
9. Calculate the percentage increase or decrease in length of the potato chips in each of the solutions.
10. Plot a graph of the change in length (per cent) against molarity.

RESULT
1. From the graph determine the molarity of the solution at which there is no change in length.

2. Record this value and from a set of tables determine the osmotic potential of this sucrose solution.

CONCLUSION
The water potential of the potato tuber is equal to the osmotic potential of this solution, which is the potential at which there is no change in length of the potato chip.

7.2

Figure 7.1 shows two chambers, A and B, separated by a membrane that only allows the passage of water across it. Tubes A and B are open to air. The water potential of pure water is zero.

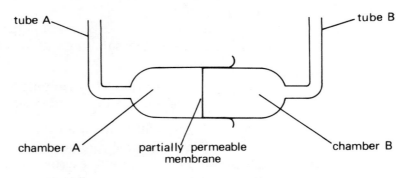

Fig. 7.1

(a) In what units should water potential be measured? ... pascals (k or M) ....

(b) If the two chambers are filled with pure water to the same level in tubes A and B, how will water molecules behave in relation to the membrane?
    Equal numbers of water molecules will pass randomly in either direction across the membrane.

(c) If chamber A and tube A contain pure water and chamber B and tube B contain a salt solution, and if the levels in the tubes are equal, how will water molecules behave in relation to the membrane?
Water molecules will move from chamber A (high ψ) to chamber B (low ψ). A few water molecules from B will pass towards A but net movement of water molecules will be from A to B.

.... **[4]**
[AEB]

7.3

Figure 7.2 shows a simplified view of the vascular system in the flowering plant.

88

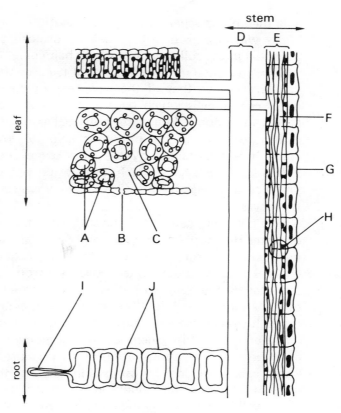

Fig. 7.2

(a) *Identify the structures labelled A to J in Fig. 7.2.*   [4]

A vacuoles of mesophyll cells   F sieve tube
B stoma   G companion cell
C intercellular air space   H sieve plate
D xylem   I root hair
E phloem   J cortex parenchyma cells

(b) *Describe BRIEFLY how raw materials (other than water) enter a flowering plant.*   [2]

Carbon dioxide enters a flowering plant by diffusion through the stomata of the leaf. Mineral salts are taken into plant root hairs from the soil solution as ions by passive diffusion and by active transport.

(c) *Describe the mass-flow theory of how organic substances are transported in the flowering plant.*   [3]

[NISEC]

The information required to answer this part of the question is provided in section 7.1(d), translocation of organic solutes.

7.4

Describe the mechanisms by which water passes through a plant from the soil to the atmosphere, showing how the various cells along its path are adapted for their function. [18]

[UCLES]

Water enters the root system through the large surface area presented by the villi-like root hairs. The cell membrane of the root hair acts as a differentially permeable membrane separating the soil solution from the cell solution in the vacuole. Water will pass from soil to cell by osmosis as long as there is a higher water potential in the soil than in the root hair cell.

As water enters the root hair cell it raises the water potential of the piliferous cell layer above that of the neighbouring cortex cells and water passes into the outer cortex cells which have a lower water potential. Water passes across the cortex in response to a water potential gradient which exists across the root from higher potential in the piliferous layer to lower potential in the xylem. The gradient is maintained by the presence of osmotically active materials in the xylem sap which has a much lower water potential than the soil solution and the presence of a negative pressure in the xylem caused by evaporation of water from the leaves.

The parenchyma cells of the cortex are tightly packed and facilitate movement of water by apoplast, symplast and vacuolar pathways. The symplast pathway is maintained by the plasmodesmata of adjacent cells and the tight packing of the cortical cells facilitates water movement from vacuole to vacuole. Free movement of water by the apoplast pathway is prevented by the waterproof Casparian strip of the endodermal cells. At this point in the passage of water the water must pass through the cell membrane of the endodermal cells.

Water enters the xylem by the combined effects of root pressure and the cohesive forces of transpiration. Root pressure is developed by the movement of water into the xylem by osmosis in response to the active secretion of salts and other solutes into the xylem sap which creates a lower water potential than that of the cells of the cortex and endodermis.

The evaporation of water from the surfaces of leaf mesophyll cells reduces their water potential. Water from the xylem vessels in the leaves replaces this lost water. This creates a negative pressure or tension in the xylem vessels in the leaves. The tension is transmitted downwards through the xylem because of the cohesive properties of continuous columns of water in the xylem vessels. The adhesive forces of water in the narrow lumens of the dead xylem vessels aid in the ascent of water. The walls of the vessels are lignified and have high tensile strength, thus preventing the tubes from collapsing when transporting water under tension.

The irregular packing of the leaf mesophyll cells creates intercellular spaces into which water vapour evaporates before diffusing out of the stomata along a diffusion gradient from leaf to air. Water loss through the stomata is controlled by the size of the stomatal apertures which is regulated by the guard cells which respond to the availability of light and water.

7.5

(a) *In the space below, make a labelled drawing of a stoma and adjacent cells as seen in surface view under the high power of a light microscope.*

[4]

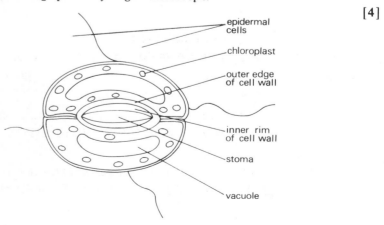

90

(b) Describe the way in which a stoma opens and closes.

As light intensity increases at dawn, K⁺ ions and associated organic anions are pumped into guard cells. This increases the solute concentration of the vacuole sap and decreases its water potential and water enters the vacuole sap by osmosis. Differential thickening of guard cell walls opposing increasing turgor pressure opens the stoma. As light intensity decreases, the reverse happens.

[4]

[L]

7.6

Figure 7.3 illustrates a piece of apparatus commonly used in plant physiology.

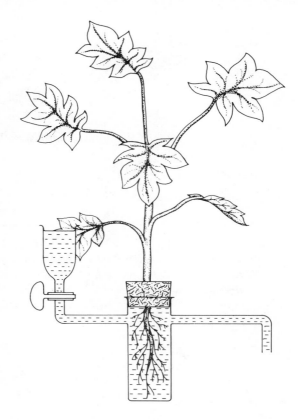

Fig. 7.3

(a) State precisely what it could be used for.

measuring the rate of water uptake by plant roots

(b) What readings would be taken and how could quantitative results be obtained?

Measure the diameter of the capillary tubing and the distance moved by an air bubble along it in a given length of time. By

continued

multiplying these values together the rate of water uptake can be calculated. Repeat to obtain mean.

(c) How would changes in

(i) relative humidity and

(ii) temperature affect such readings? Offer explanations for your answers.

(i) As humidity increases the diffusion gradient of water between the atmosphere and the sub-stomatal space decreases reducing the rate of water uptake. A reduction in humidity has the opposite effect.

(ii) As temperature increases the rate of diffusion of water into the atmosphere increases. This will increase the rate of water uptake. A reduction in temperature will have the opposite effect.

[6]
[UCLES]

7.7

(a) State three factors that directly influence the rate of transpiration from the leaves of a flowering plant.

First factor . . . . *light* . . . .

Second factor . . . . *temperature* . . . .

Third factor . . . . *humidity* . . . .

[3]

(b) Using the axes below, draw a curve to indicate the changes you would expect in the rate of transpiration of a herbaceous flowering plant over a 24-hour period during hot, dry, sunny weather.

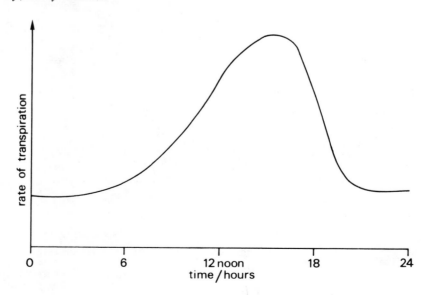

[4]

(c) Explain the significance of the shape of the curve you have drawn in (b).

During the darkness some water loss occurs through the cuticle. This rate of water loss is fairly constant. Following dawn, light intensity increases, the stomata open and water is lost by evaporation. As temperature rises towards early afternoon the rate of transpiration increases to a peak before falling off rapidly as light intensity and temperature decrease at dusk.

.... **[4]**

[L]

7.8

(a) Figure 7.4 represents the blood circulation in a fish.

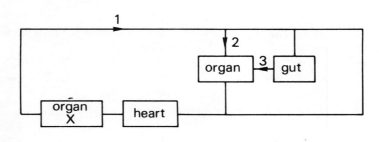

Fig. 7.4

(i) Name the organ labelled X.

gills

.... **[1]**

(ii) By inserting arrows, show the direction of blood flow through the blood vessels labelled 1, 2 and 3. **[2]**

(b) Complete the table below which compares three blood circulatory systems.

	Number of chambers		Form of circulatory system
	Atria (auricles)	Ventricles	
Mammal	2	2	Complete double
Fish	1	1	*single*
Frog	2	1	*partial double*

(c) Figure 7.5 (overleaf) shows a human heart. **[2]**

(i) On the diagram, mark with the symbol 'CO_2' the artery which carries blood with the highest concentration of carbon dioxide. **[1]**

(ii) Mark on the diagram using the letter 'M' the position of the pacemaker. **[1]**

continued

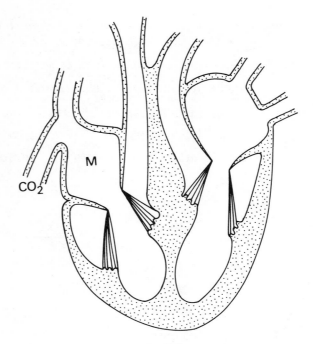

Fig. 7.5

(iii) Name the part of the brain which controls the pacemaker.

medulla oblongata

. . . . **[1]**

(iv) A pilot's rate of heartbeat during landing increases considerably. Name the hormone which causes this.

adrenaline

. . . . **[1]**

[SEB]

7.9

Figure 7.6 shows some of the structures visible in a ventral view of a section through a mammalian heart.

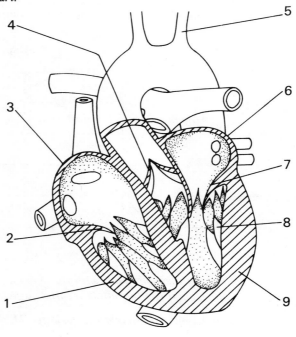

Fig. 7.6

(a) *Name the parts numbered 2, 3, 4, 5, 7 and 8.*

2 ....tricuspid valve..

3 ....right atrium...

4 ....semilunar valve.......................................

5 ....left carotid artery....................................

7 ....mitral valve..

8 ....papillary muscle......................................

[6]

(b) *What is the significance of the difference in thickness between (i) 1 and 9 and (ii) 6 and 9?*

(i) *1 and 9* ...thickness 1 required only to pump blood to lungs. Greater force needs to be produced by 9 to pump blood around body.

....

(ii) *6 and 9* ......thickness 6 only needed to pump blood into ventricle hence less force needed than produced by 9.

....

[2]

(c) *When the valve numbered 4 is closed, indicate by a tick in the relevant column of the following table the condition of the numbered structures.*

Structure	Condition	
	Contracting	Relaxing
1		✓
3	✓	
8		✓
9		✓
	Opened	Closed
2	✓	

[5]

(d) *Suggest the probable response of the heart to each of the following changes.*

(i) *Stimulation by the sympathetic nervous system.*
....increases rate of contraction of heart muscle.........

(ii) *Increased carbon dioxide concentration in the blood.*
....increases rate of contraction of heart muscle.........

[2]

[L]

7.10

Figure 7.7 shows a vertical section through the heart of a mammal. The arrows show the direction of blood flow.

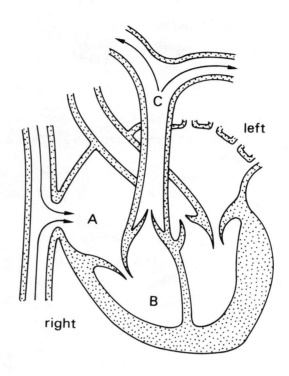

Fig. 7.7

The graphs (Fig. 7.8) show changes in blood-pressure in A, B and C together with the simultaneous changes in volume of cavity B as the heart beats. Points along the graph lines are marked with crosses and numbered so that they may be easily referred to.

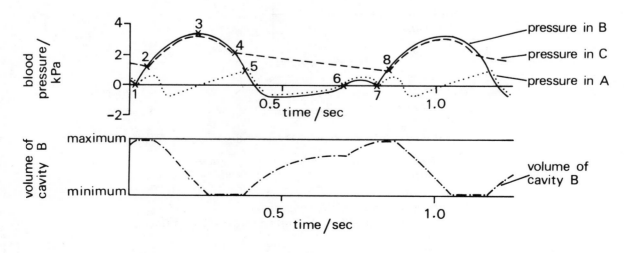

Fig. 7.8

96

(a) Describe the route taken by blood through the right side of this heart. [6]

(b) Between which points on the graph will the valve between
  (i) A and B be closed? [1]
  (ii) B and C be closed? [1]

(c) What happens to the flow of blood in A when the pressure in B is negative? [2]

(d) (i) How is pressure in B increased between points 1 and 3? [5]
  (ii) Explain why the volume of cavity B remains the same between points 1 and 2 even though the pressure is increasing. [2]

(e) How do you explain the surge of pressure in A and B between points 6 and 7? [3]
[AEB]

(a) Blood enters the right atrium from the superior and inferior venae cavae while the tricuspid valve is closed. When the atrium is completely full, atrial systole occurs and forces blood through the tricuspid valve into the right ventricle. As blood enters the ventricle the venae cavae sphincters and pulmonary artery valves close. When the ventricle is completely full, ventricular systole occurs, the tricuspid valve closes and the pulmonary artery valves open, allowing blood to leave the right venticle. The cycle then repeats.

(b) (i) 1 to 5
  (ii) 4 to 8

(c) While the pressure in **B** is negative the right atrium (**A**) gradually fills up with blood from the venae cavae. Blood passes into the ventricle.

(d) (i) Between 1 and 3 the muscles in the wall of **B** contract. This is ventricular systole. The pressure initially increases because the volume of the ventricle decreases without loss of blood. Only when the semilunar valves of the pulmonary artery open can the pressure begin to decrease, but this is offset by the increased force produced as the ventricle muscle contracts.
  (ii) Between 1 and 2 both sets of valves are closed.

(e) Points 6 and 7 correspond with the period of atrial systole. This increases pressure within A due to the reduction in volume of A and increases the pressure in B due to the flow of blood into B through the tricuspid valve.

7.11

The oxygen dissociation curves (Fig. 7.9) represent the relationship between the partial pressure of oxygen and the percentage oxygen saturation of two respiratory pigments. Curve A shows the response of myoglobin in muscle and curves B, C and D the response of haemoglobin in the blood at three different partial pressures of carbon dioxide.

KEY
A = myoglobin in muscle
B = haemoglobin at CO_2 partial pressures 2666 Pa† (20 mmHg)
C = haemoglobin at CO_2 partial pressures 5332 Pa (40 mmHg)
D = haemoglobin at CO_2 partial pressures 10 664 Pa (80 mmHg)

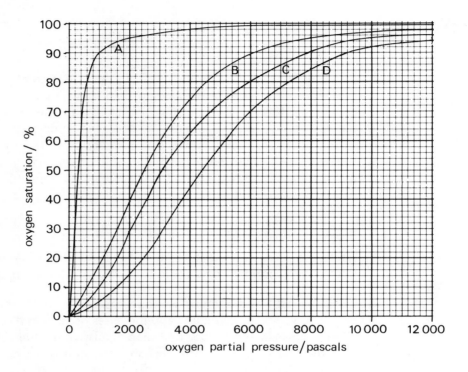

Fig. 7.9

†A pascal (Pa) is a unit of pressure. A pressure of 100 000 pascals is approximately equal to atmospheric pressure (760 mm Hg).

(a) For curve B, briefly describe the effect that increasing partial pressure of oxygen has on the saturation of the respiratory pigment.

Uptake of oxygen by the pigment is rapid as the partial pressure rises to 6000 pascals. After a saturation of 90% the oxygen uptake is reduced.

[2]

(b) Over which range of partial pressures of oxygen does the most rapid reaction with haemoglobin occur for (i) curve B and (ii) curve D?

(i) curve B ... *0 – 3000 pascals*

(ii) curve D ... *3000 – 5000 pascals*

[2]

(c) *Where in the body of a mammal is the partial pressure of carbon dioxide likely to be high, as in curve D?*

liver

[1]

(d) *If the blood when fully saturated with oxygen is able to carry 100 cm³ (ml) of oxygen per dm³ (litre), calculate the volume of oxygen released per dm³ (litre) from the blood when blood that is 90 per cent saturated flows into a tissue where the partial pressure of oxygen is 4000 Pa (30 mmHg) and that of carbon dioxide is 5333 Pa (40 mmHg).*

When 100% saturated the blood carries 100 cm³ of oxygen, ∴ 90% saturated carries 90 cm³. From curve C at 4000 Pa blood is 62% saturated, ∴ carries 62 cm³. 90−62 = 28 cm³ per litre.

[3]

(e) *Suggest reasons for the differences in curve A and curve D.*

Myoglobin releases oxygen at low partial pressures of oxygen, whereas haemoglobin releases oxygen steadily across a range of partial pressures. These are adaptations to functions in the body.

[3]

(f) *From curves B, C and D it is clear that increasing partial pressure of carbon dioxide affects the ability of the respiratory pigment to combine with oxygen. Give the name for this phenomenon.*

Bohr shift

[1]

(g) *Curve A is similar to a curve obtained when investigating the oxygen-carrying capacity of the respiratory pigment in an aquatic worm which burrows in mud. Explain how this curve indicates the worm's adaptation to its environment.*

The pigment saturates with oxygen rapidly with increasing oxygen partial pressures. At about 1500 Pa it is almost fully saturated. This represents conditions when mud is aerated. As oxygen is required by the tissues there is a rapid release of oxygen by the pigment.

[3]
[L]

7.12

Overleaf is a table for recording tests for blood transfusions in the ABO system.

(a) *In the appropriate boxes place a tick where agglutination does **not** occur and a cross where agglutination would occur.*

continued

99

	Recipients			
Donors	*Group A*	*Group B*	*Group AB*	*Group O*
Group A	✓	✗	✓	✗
Group B	✗	✓	✓	✗
Group AB	✗	✗	✓	✗
Group O	✓	✓	✓	✓

[8]

(b) *The properties of these blood groups are controlled by three allelic genes:*

I^A *for antigen A;*
I^B *for antigen B;*
i *for no antigen.*

    (i) State the genotype of group A when it is:

(1) homozygous; .... $I^A I^A$ ....

(2) heterozygous; .... $I^A i$ ....

[2]

    (ii) State the possible genotypes of group B when it is:

(1) homozygous; .... $I^B I^B$ ....

(2) heterozygous; .... $I^B i$ ....

[2]

    (iii) State the genotype of group O. .. ii .... [1]

    (iv) How would you describe the type of dominance of genes I^A and I^B?

        codominance

.... [1]

    (v) If a group O man married a group AB woman what blood groups could their children have? ....

.... Group A and group B ....

.... [2]

[OLE]

100

8 Plant Co-ordination

Tropism The growth movement of part of a plant in response to, and directed by, an external stimulus. The response is either positive or negative, depending on whether growth is towards or away from the stimulus respectively.

Taxis The movement of a whole cell or organism in response to, and directed by, an external stimulus. It may be positive or negative as for tropisms.

Nasty A non-directional movement of part of a plant in response to an external stimulus. The nature of the movement is determined by the structure of the responding organ.

Bioassay An experiment in which the amount of a substance is found by measuring its effects on a biological system.

Apical dominance The presence of a growing apical bud inhibits growth of lateral buds. It also includes the suppression of lateral root growth by growth of the main root.

Abscission The organized shedding of parts of a plant.

Photomorphogenesis The effect of light upon plant development.

Photoperiodism The response of organisms to varying periods of light and dark in the 24-hour day.

8.1 Plant Movements

Plant movements occur in response to external stimuli. Motile algae display **tactic** movements, the whole organism moving as directed by the stimulus. **Nastic** movements include the sleep movements (**nyctinasty**) of some flowers where they open and close in response to light (**photonasty**) and temperature (**thermonasty**).

8.2 Plant Growth

The orderly growth and development of plants is chemically controlled by growth regulator substances. These are produced in minute quantities in specific parts of the plant and transported to other regions where they modify the pattern of growth. The process invariably involves a permanent change in the structure of the plant. Auxins and gibberellins are associated with cell enlargement and differentiation, cytokinins with cell division, abscisic acid with resting states and ethene with ageing.

Plants respond by differential growth towards (*positive*) or away from (*negative*) a variety of stimuli. These include light (**phototropism**), gravity (**geotropism**), water (**hydrotropism**), touch (**thigmotropism**), chemicals (**chemotropism**), and temperature (**thermotropism**). Tropic responses are controlled by auxins.

8.3   Plant Growth Regulator Substances

(a) Auxins

In 1934 auxin was identified as **indole acetic acid (IAA)**. Auxins are manufactured in shoot and root apices, young leaves, buds and seeds. They diffuse in one direction only away from their source to where the response takes place. If shoot tips are subjected to unilateral light, auxin migrates laterally away from light. Therefore more auxin accumulates on the shaded side. Larger quantities of auxin cause greater cell elongation and the shoot consequently curves towards the light. This is called **positive phototropism**. When the shoot points towards the light the response ceases as the light is no longer unilateral. As auxin migrates down from the tip it is progressively inactivated and broken down by enzymes.

Auxins inhibit growth in lateral buds, thus confining growth to the apex of the plant (**apical dominance**). This is an example of **correlation**—one part of a plant controlling the development of another via the influence of a growth substance.

The same quantity of auxin that promotes growth in the shoot inhibits growth in the main root. In a laterally placed root, auxin is translocated downwards to the lower side of the root. Here the greater auxin concentration inhibits cell enlargement. Therefore the cells on the upper side grow faster and the root curves towards gravity. This is called **positive geotropism**. Removal of the root cap inhibits the geotropic response.

The 'starch-statolith' hypothesis states that mobile starch grains in the root cap cells act as statoliths and migrate to the lower sides of the cells in response to gravity. This in some way affects auxin distribution. In a number of plants abscisic acid, ethene and/or gibberellin control gravity responses of roots rather than auxins.

Auxins stimulate proton secretion causing an increase in acidity outside the cell. This stimulates an enzyme-controlled process which loosens the bonds between the cellulose microfibrils of the cell wall. This decreases wall pressure and allows more water to enter. As endosmosis occurs the turgor pressure generated causes extension of the cell walls.

Auxins also induce the formation of adventitious roots. They inhibit abscission of leaves and fruits. Auxins produced by pollen tubes and seeds stimulate growth of the ovary wall. The different responses of different organs to auxins is thought to be due to differences in sensitivity of plant tissues. Commercially, auxins are used to help fruit set, and stimulate fruit development by parthenocarpy. Properly applied, they will initiate root development of cuttings and inhibit the sprouting of potatoes. Auxins are also employed as weedkillers.

(b) Gibberellins

Gibberellins are abundant in young expanding organs and are translocated up and down the plant in the phloem or xylem. Their principal effect is to elongate stems through cell elongation. They also break the dormancy of buds, aid setting of fruit and affect flowering. They promote seed germination by stimulating the production of enzymes which direct the mobilisation of food reserves which are then used for growth by the embryo.

In cell elongation, gibberellin action depends upon the presence of auxins. In cereal grains gibberellins may stimulate protein synthesis and be involved with the regulation of gene activity during differentiation. Commercially, gibberellins are produced from fungal cultures and used for fruit setting, and growing seedless grapes through parthenocarpy.

(c) Cytokinins

Cytokinins are chemically similar to the base adenine. They increase the rate of cell division in fruits and seeds, stimulate bud development and increase the rate of cell enlargement in leaves. However, they only promote cell division by interacting with auxins (and most probably gibberellins as well). They may also delay the ageing process in leaves. Commercially, cytokinins are used to prolong the life of fresh leaf-crops such as lettuce, and flowers.

(d) Abscisic acid (ABA)

Abscisic acid (ABA) is made in roots, leaves, stems, fruits and seeds. It moves through the plant in the phloem and diffuses into the root from the root cap. ABA is a major inhibitor of plant growth and is antagonistic to the growth promoters gibberellin and cytokinin. Its main effects are to inhibit cell elongation and seed germination, promote dormancy in buds, and induce leaf abscission. ABA is associated with stressful situations, especially drought. In tomatoes under such conditions its concentration is fifty times more than normal. ABA may also stimulate stomatal closure. Commercially, ABA is sprayed on fruit crops to control fruit drop. This means that fruit picking can take place over a predetermined short period.

During abscission, an abscission layer is formed by the breakdown of cells in the abscission zone at the base of the organ. The fruit or leaf is shed when its point of attachment to the parent plant is broken mechanically, e.g. by wind. A protective layer below the abscission layer prevents infection or desiccation and the vascular tissue is sealed. Abscission is probably controlled by interactions between auxin (which inhibits) and ABA (which promotes). As the leaf approaches abscission its auxin concentration declines. However, once abscission begins, auxin accelerates the process.

(e) Ethene

Ethene is a product of metabolism of most plant organs. It may act as a growth inhibitor and can promote abscission of fruits and leaves by stimulating the production of enzymes involved in these processes. Commercially it is used to stimulate ripening of tomatoes and citrus fruits.

(f) Regulator Substances: General

While the effects of these growth substances have been treated separately, it must be emphasized that generally they work by interacting with one another. Where the combined effect of two or more substances works to influence growth this is called synergism. Antagonism occurs where two substances exert opposite effects on the same process. Table 8.1 indicates several situations where such interactions occur though this is by no means a complete list.

Table 8.1

Process affected	Auxin	Gibberellin	Cytokinin	ABA	Ethene
Stem growth	Promotes cell enlargement Promotes cell division at cambium	Promotes in presence of auxin	Promotes cell division in apical meristem and cambium	Inhibits especially during physiological stress e.g. drought	Inhibits especially during physiological stress e.g. drought
Root growth	Promotes at low concentrations; inhibits at high concentrations	Inactive	Inactive or inhibits primary root growth	Inhibits	Inhibits.
Root initiation	Promotes growth of roots from cuttings	Inhibits	Promotes lateral root growth	–	–
Fruit growth	Promotes	Promotes	Promotes	–	–
Apical dominance	Promotes	Enhances auxin action	Antagonistic to auxins	–	–
Abscission	Inhibits	Inactive	Inactive	Promotes	–

8.4   Flowering (a Photoperiodic Response)

The onset of flowering is regulated by variations in daylength. This ensures that plants of the same species flower at the same time, thus promoting the chances of cross pollination and cross fertilization. Light is absorbed in the leaves by a blue–green photoreceptor pigment called **phytochrome**. It exists in two interconvertible forms, P_{730} and P_{660}. Absorption of light by one form of phytochrome converts it rapidly to the other form. When this occurs the stimulus is transmitted to the flowering apices by the phloem and flowering is stimulated. The transmitting agent is thought to be a hormone which has been hypothetically named 'florigen'.

Day-neutral plants, e.g. tomato and dandelion, flower independently of the daylength however variable it might be.

Long-day plants, e.g. onion and potato, flower only if they are exposed to more than ten hours of light during each 24-hour period, e.g. during the summer months in temperate regions. They flower when the presence of red light (a long period of sunlight) causes sufficient accumulation of P_{730} and a low level of P_{660}.

Short-day plants, e.g. primrose and strawberry, flower only if they are exposed to less than 12 hours' light during each 24-hour period. They flower when a long period of darkness or far-red light causes sufficient accumulation of P_{660} and a low level of P_{730}:

Other phytochrome responses include folding and opening of leaves during night and day respectively, and in conjunction with ABA, leaf abscission and initiation of dormancy.

8.5   Vernalization

Before some plants can flower they must be subjected to low temperatures (around 4 °C). Flowering is then subsequently controlled by photoperiodism. The cold stimulus is perceived by the mature stem apex or the embryo of the seed. During vernalization the level of gibberellins increases and it is thought that they transmit the stimulus throughout the plant. Photoperiodism and vernalization synchronize the reproductive behaviour of plants so that it takes place at favourable times of the year.

Winter bud formation is a photoperiodic response to the shortening days of autumn. The stimulus, perceived by the leaves, causes ABA to accumulate. ABA migrates to the meristems and inhibits further growth. Many buds must experience a period of cold before dormancy is broken (bud break). Likewise some seeds need to be chilled (stratification) after imbibing water, before they can germinate.

8.6   Worked Examples

8.1

(a) *Name two meristematic regions in a plant shoot.*

(i) ....*cambium*.................. (ii) ....*shoot apex*.......... [2]

(b) *Name two plant growth substances (hormones). For each give one function.*

	Name of hormone	*Function*
(i)	*indoleacetic acid*	*stimulates cell elongation in shoots* [2]
(ii)	*gibberellic acid*	*stimulates cell elongation in young shoots* [2]

(c) (i) *What does the term short-day plant mean?*
A plant that will only flower if exposed to less than 12 hours of light in each 24 hour period. [2]

(ii) *What mechanism controls the short-day response?* ....
phytochrome activation [2]

[OLE]

8.2

Figure 8.1 refers to an experiment with oat coleoptiles.

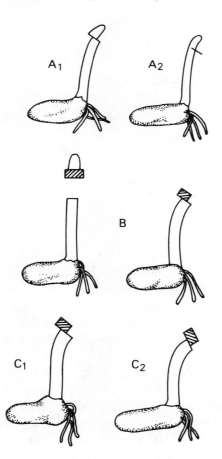

A₁ shows the appearance of an oat coleoptile some time after the tip has been cut off and replaced asymmetrically.

A₂ shows a similar coleoptile some time after a razor blade has been placed halfway through the coleoptile below the tip.

B shows an oat coleoptile from which the severed tip has been placed on an agar block for two hours. The agar block has then been placed on the coleoptile stump asymmetrically.

C₁ and C₂ are oat coleoptiles in which the tips have been replaced with agar blocks placed asymmetrically. C₁ contains a low concentration of auxin and C₂ a higher concentration of auxin.

Fig. 8.1

(a) *What changes occur in the coleoptiles to produce the effects shown in (i)A_1 and (ii) A_2?*

(i) A_1 *increased cell elongation on side bearing the tip. Less elongation on the other side.*

(ii) A_2 *increased cell elongation on side away from blade. Less elongation on the other side.*

[4]

(b) *What conclusions can you draw from the results of experiments (i) B and (ii) C_1 and C_2?*

(i) B *tip produces growth stimulating substances which diffuse into the agar block and then into the cut coleoptile where they stimulate cell elongation.*

(ii) C_1 and C_2 *the degree of cell elongation is proportional to the concentration of auxin.*

[4]

(c) *State three precautions you would take in conducting an experiment of this kind and, in each case, given one reason.*

	Precaution	Reason
(i)	use same batch of seed	ensure genetic and physiological similarity of seed
(ii)	use adequate controls in A, B and C	ensure that responses shown only occur in experimental situations
(iii)	all other variables must be kept constant	ensure that no other extraneous factor influences the response

[6]

(d) *How may the concentration of naturally occurring plant hormones be measured?*

By use of a technique called biological assay. The angle of curvature of a coleoptile such as in C_1 and C_2 can be measured. This value can be compared with angles of curvature using known concentrations of artificially produced hormones. These angles are proportional to auxin concentrations.

[4]
[L]

8.3

Name the behavioural response shown by each of the following.

(a) *A root growing towards gravity.*

positive geotropism

(b) Chlamydomonas *(a motile alga) swimming towards light.*

positive phototaxis

(c) *The leaflets of* Mimosa *drooping at night.*

photonasty

(d) *The leaf of a Venus fly trap plant closing when a fly lands on it.*

haptotropism   thigmonastic

(e) *A plumule growing upwards through the soil.*

negative geotropism , Positive phototropism

(f) *The male gamete of a fern swimming towards an archegonium.*

positive chemotaxis

(g) *A runner bean twining around its support.*

positive thigmotropism

(h) *A coleoptile growing towards unilateral light.*

positive phototropism

(i) *A tulip flower closing when the temperature drops.*

negative thermonasty

[9]
[L]

8.4

Complete the following table:

Name of hormone	Site of production	Normal effect
Insulin	islets of Langerhans	stimulates uptake of glucose
Ethene (ethylene)	fruits and seeds	ripening of fruit
Growth hormone	anterior pituitary	control of skeletal growth
Indoleacetic acid (IAA)	shoot apex	causes bending of coleoptile
Progesterone	ovary	prevents ovulation
Oxytocin	posterior pituitary	release of milk reflex
Secretin	walls of duodenum	pancreatic juice release
Abscisic Acid	leaf petiole	leaf fall

[5]
[OLE]

8.5

The table shows the effects of applying different concentrations of auxin to the shoots and roots of cereal seedlings. A positive value indicates increased elongation and a negative value indicates reduced elongation (compared with elongation in control plants in the same time interval).

Auxin concentration (parts per million)	Elongation relative to control (mm)	
	Shoot	Root
10^{-6}	0	+2
10^{-5}	0	+4
10^{-4}	+1	+12
10^{-3}	+4	+8
10^{-2}	+8	−2
10^{-1}	+36	−24
1	+60	−40
10	+32	−40
100	−24	−40

(a) Express these results graphically on the same set of axes. [4]

(b) Explain how these results may be related to the geotropic responses shown by stems and roots. [6]

(c) Describe **two** other ways in which auxins can affect plant growth and development. [4]

(d) Briefly compare the roles of auxins, gibberellins and cytokinins (e.g. kinetin) in plants. [3]

(e) Synthetic auxins are easily manufactured and are often used as selective herbicides. Suggest how these may operate if used in a lawn weedkiller. [3]

[AEB]

(a)

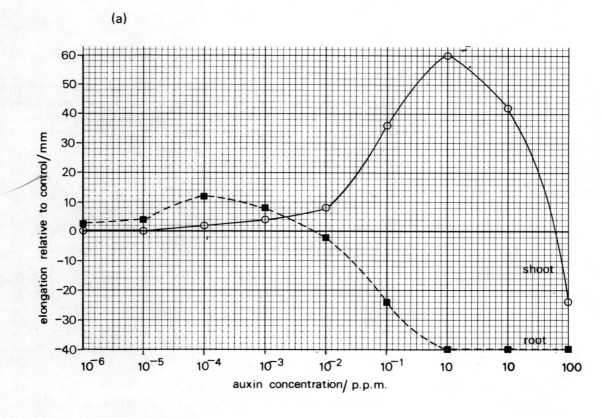

(b) Auxins are produced and released by the tips of stems and the tips of roots. Their effects on stems and roots are different. Auxins cause elongation of cells in roots at relatively lower concentrations than in stems. At concentrations of 10^{-4} and 10^{-3} p.p.m., auxin stimulates more elongation in roots than in stems but at concentrations greater than this, up to 100 p.p.m., the response to auxin changes and cell elongation in stems is promoted while cell elongation in roots is inhibited. Superimposed on these differential rates of growth by cell elongation is the geotropic response produced by the movement downwards of auxin in response to gravity so as to accumulate on the lower side of the developing root and stem. The increased concentration of auxin on the lower side of the stem stimulates cell elongation on this side, which results in upward growth of the stem while, in the same concentrations, it inhibits growth of the root on the lower surface. Thus at concentrations of, say, 1 p.p.m., auxin produces differential growth in stems and roots, leading to the geotropic responses of upward growth of stems and downward growth of roots.

(c) Auxins can inhibit growth in lateral buds in stems and induce the formation of adventitious roots in root systems.

(d) Both auxins and gibberellins promote stem elongation by stimulating cell enlargement, whereas cytokinins promote stem elongation by stimulating cell division. Many of the effects of both gibberellins and cytokinins depend upon the presence of auxins. The opposite is not the case.

(e) Many auxins, in particular the phenoxyacetic acids 2,4–D and MCPA, can selectively destroy dicotyledons which exist as weeds within a 'stand' of monocotyledons, as for example on a lawn. Since dicotyledons have broader leaves than monocotyledons they take up more weedkiller. The auxins cause rapid growth of the dicotyledons which grow in a distorted way and eventually outgrow themselves and die.

8.6

Roots are positively geotropic. The generally accepted hypothesis to explain geotropism is that auxin accumulates on the lower side of horizontally placed roots and inhibits cell elongation there, thus causing a downward curvature. Modern investigations, however, show no such accumulation of endogenous auxin on the lower side.

Figure 8.2 illustrates results obtained by Pilet using root tips of maize (Zea mais). He removed the entire root caps and part of the meristem from some roots (i, ii and iii) and half of the root caps and part of the meristem from others (iv, v and vi).

(a) What does the fact that those roots without root caps hardly curved at all indicate?
    Meristems have little geotropic activity in the absence of root caps which are necessary for producing the response.

(b) How would you explain that those with half their root caps intact (iv, v and vi) consistently curved in the direction of the remaining cells?
    An inhibitor substance produced by the root cap diffuses backwards, is relatively unaffected by gravity and prevents elongation.

(c) From this evidence, suggest another hypothesis to explain geotropism.
    The geotropic response may involve another growth regulator substance other than auxin.

[6]
[UCLES]

110

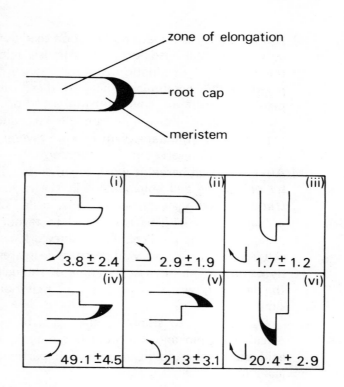

Fig. 8.2 Geotropic curvature in maize root tips as affected by removal of root cap and meristematic cells of the root apex. Direction of curvature is indicated by arrows, and numbers represent degrees of curvature.

8.7

In the table below, state three differences between plant hormones and animal hormones.

	Plant hormones	Animal hormones
(i)	slow transportation	rapid transportation in blood stream
(ii)	transported usually by diffusion	transported in blood plasma
(iii)	no specific sites of synthesis	synthesized in endocrine glands

[3]
[L (adapted)]

8.8

Discuss the role of growth substances in (a) seed dormancy, (b) seedling growth and (c) leaf fall. [4, 10, 4]
[UCLES]

There are five main groups of growth regulator substances: auxins, gibberellins, cytokinins, abscisic acid and ethene.

(a) Auxins and ethene have no effect on seed dormancy, whereas gibberellins and cytokinins break seed dormancy. Abscisic acid promotes seed dormancy but as the level falls germination proceeds. Once dormancy has been broken as a result of the metabolic activity of these substances, germination will begin, provided optimum environmental conditions exist. Certain seeds, notably cereals, secrete gibberellins from the aleurone layer at an early stage in germination. This stimulates synthesis of several enzymes involved in the breakdown of food reserves in the endosperm.

(b) Auxins and gibberellins work together to stimulate stem growth in seedlings by promoting cell enlargement in the region behind the stem apex. Auxins stimulate hydrogen ion secretion by cells in this region which reduces the pH outside the cells. This is believed to favour the activity of enzymes which break down the rigid cellulose framework of the cell. As the 'plasticity' of the cell wall increases, cell wall pressure decreases, allowing water to enter the cell by osmosis. The increased volume of the cell leads to cell elongation prior to the laying down of new cell wall material. Gibberellins may operate in a similar way.

Cytokinins also stimulate stem growth by promoting cell division in the apical meristem and cambium but only in the presence of auxins. Abscisic acid and ethene inhibit stem growth, particularly during periods of physiological stress.

Very low concentrations of auxins promote root growth. At higher concentrations they have an inhibitory effect, as do cytokinins, abscisic acid and ethene. It is the interplay of these various effects on stem and root growth in seedlings that produces phototropic and geotropic responses.

At a later stage in seedling growth gibberellins and cytokinins promote leaf growth.

(c) Leaf fall (abscission) is determined by the interaction of several environmental factors which affect the balance of some of the main growth regulator substances. Auxins usually inhibit abscission. Abscisic acid promotes the development of an abscission layer but the mode of action is not at all clear. Gibberellins and cytokinins do not appear to have any effect on abscission.

9 Animal Co-ordination

Irritability   The ability to respond to stimuli.

Action potential   A rapid change in membrane potential resulting in the transmission of a nerve impulse.

Nerve impulse   An electrochemical wave of depolarization transmitted along a nerve fibre.

Sensory neurone   A neurone that carries impulses from a receptor to the central nervous system.

Motor neurone   A neurone that transmits nerve impulses from the central nervous system to effectors.

Nerve   A bundle of nerve fibres enclosed by a connective tissue sheath.

Dendrite   A branched nerve fibre that conducts impulses towards the cell body.

Axon   An unbranched elongated nerve fibre that carries impulses away from the cell body.

Chemoreceptor   A cell or organ that detects substances according to their chemical structure, e.g. taste and smell.

Mechanoreceptor   A cell or organ that detects mechanical stimuli, e.g. touch, pressure, hearing, balance.

Effector   A structure capable of producing a response to stimuli.

Conditioned reflex   A learned response to a stimulus. The brain is involved but with continued repetition the response becomes involuntary.

Hormone   A complex organic substance of variable composition secreted in minute amounts by a gland, it passes into the bloodstream and affects target cells or tissues at some distance away from its point of origin.

9.1   Co-ordinating Systems

The principal co-ordinating systems of the body are the nervous and endocrine systems. Functionally, the nervous system is capable of producing a rapid response mediated by chemical and electrical transmission acting on muscles. Hormones are chemicals which are transmitted through the bloodstream. They produce slower, longer-term responses and work by altering the metabolic activity of their target cells.

9.2   Nervous Communication

The vertebrate nervous system consists of a **central nervous system (CNS)** (the brain and spinal cord), and a **peripheral nervous system** (nerve fibres which connect receptors and effectors to the CNS). Changes in the external or internal environment, called **stimuli**, are detected by **sensory receptors**. This information is passed along conductile **neurones** in the form of **nerve impulses** to the CNS. Here the information is analysed and impulses are sent to effectors which respond in an appropriate manner.

Neurones are the units of structure and function of the nervous system. They are variable in shape and size and are responsible for conducting nerve impulses throughout the body. A typical **motor neurone** consists of many branched, hair-like **dendrites**. These are in synaptic contact with adjacent neurones and receive and translate incoming stimuli into nerve impulses which are conveyed to the cell body. The **cell body** contains the nucleus. A single, long cytoplasmic extension, the **axon**, or **nerve fibre**, runs from the cell body. The axon is enclosed by a cell membrane. Surrounding this is the fatty **myelin sheath** formed by **Schwann cells**. This in turn is bounded by the **neurilemma**. The myelin sheath protects and provides electrical insulation for the axon and speeds up impulse transmission. It is interrupted at intervals by **nodes of Ranvier**. At the end of the axon, small, branched dendrites are present which terminate in **synaptic knobs**. These form synapses with other cells and convey impulses to neighbouring neurones by release of a chemical **neurotransmitter substance**.

In a resting axon the membrane is relatively impermeable to Na^+ ions but twenty times more permeable to K^+ ions. Also, an active **cation pump** couples expulsion of Na^+ ions from the *inside* of the axon with intake of K^+ ions from *outside* the axon. However, K^+ ions tend to leak out of the axon down the steep K^+ gradient. This creates a net negative charge inside the axon relative to its exterior. This potential difference across the membrane is of the order of -60 to -70 mV and is called the **resting potential** of the axon.

Impulses are generated when receptor cells respond to stimuli. The nature of the impulse elicited does not vary whatever the nature of the stimulus or receptor. For an **action potential** to be produced, the stimulus has to reach a certain threshold intensity. However, further increase in the strength of the stimulus does not increase the amplitude of the action potential. This is called the '**all or nothing law**'. The strength of the stimulus is passed on as a **frequency code**, a stronger stimulus generating more impulses than a weaker one.

When an action potential is produced, the axon membrane suddenly becomes more permeable to Na^+ ions, which rush into the axon, making its interior temporarily positive relative to the exterior. This is called **depolarization**. Adjacent sections of the axon are similarly affected by the action potential and the impulse travels in a wave-like manner without decrement from one end of the axon to the other. After an impulse has passed, the axon enters an **absolute refractory period** of 1 ms when it is completely incapable of transmitting any further impulses. This is followed by a **relative refractory period** of 5 to 10 ms when only high-intensity stimuli may generate an action potential. During the refractory period the resting potential is re-established. K^+ ions leave the axon in response to entry of Na^+ ions. Then the cation pump removes the Na^+ ions from the axon, completing repolarization, and the membrane again becomes impermeable to Na^+ ions. Therefore the resting potential is largely determined by K^+ ions and the action potential by Na^+ ions.

In *myelinated* fibres action potentials jump from node to node in a process called **saltatory conduction**.

Impulses pass from neurone to neurone via chemical or electrical synapses. Synapses relay impulses in one direction only. Neurones may possess several thousand synaptic links.

At a chemical synapse the **presynaptic membrane** of a synaptic knob is separated from the **postsynaptic membrane** of a dendrite or cell body by a **synaptic cleft** 20 nm wide. Impulses arriving at the knob increase the permeability of the presynaptic membrane to Ca^{2+} ions and cause the release of neurotransmitter substance into the cleft. This is called **excitation-secretion coupling**. The transmitter combines with specific receptor sites in the postsynaptic membrane causing either **depolarization** or **hyperpolarization**. In an **excitatory** synapse depolarization causes the generation of an **excitatory postsynaptic potential**. At a particular threshold this fires an action potential in the adjacent neurone. In an *inhibitory* synapse, hyperpolarization occurs, producing an inhibitory postsynaptic potential which prevents development of an action potential. In this way synapses serve to integrate and modify precisely numerous internal activities. In cholinergic fibres the neurotransmitter substance is **acetylcholine (ACh)**. Once it has produced its effect it is decomposed rapidly by **acetylcholinesterase** into choline and acetic acid. These products diffuse back into the knob and are recycled and re-used. The neurotransmitter secreted by adrenergic fibres of the sympathetic nervous system is **noradrenaline**.

Electrical synaptic transmission occurs across synaptic clefts of 2 nm or less. There is no delay in transmission time and the passage of impulses is unaffected by drugs or other substances.

A neurone may not be excited by the amount of transmitter substance secreted by one terminal knob. However, the combined amounts of transmitter substance from several closely associated knobs may cause impulse generation. This is called **spatial summation**. **Temporal summation** occurs where the combined amounts of transmitter substance caused by a series of impulses reaching a single knob cause impulse generation at the synapse. In effect, the first impulse arriving at the synapse fails to cross it but makes it easier for the passage of the second one. This is called **facilitation**. If a synapse receives a continuous stream of impulses over a long period of time the transmitter substance in the knob becomes exhausted and transmission of impulses ceases. This is called **accommodation**.

At a neuromuscular junction each muscle fibre possesses a specialized region, the **motor end plate**. Here ACh discharged from the nerve fibre attaches to the **sarcolemma** receptor sites. This causes depolarization and the formation of an **end plate potential**. Above a certain threshold this leads to action potential generation along the muscle fibre via the **T-system**, causing the muscle to contract.

(a) Voluntary Nervous System

In the forebrain the **cerebrum** controls memory, imagination, olfaction, thought, intelligence and personality. **Sensory areas** receive information from the rest of the body and **association centres** compare this with previous information. It is integrated and computed before an appropriate motor response is given by the **motor regions** of the brain. The **thalamus** is a processing and integrating centre and can distinguish between pain and pleasure. The **hypothalamus** is the main control centre of the endocrine and autonomic nervous systems. It is involved in control of sleep, feeding, drinking, body temperature and osmoregulation.

The midbrain contains **visual and auditory reflex centres** in the **corpora quadrigemina**. The **red nucleus** controls movement and posture. Nerve tracts pass through the midbrain to connect the fore- and hindbrains.

In the hindbrain the **cerebellum** co-ordinates voluntary muscular activity and the reflex control of body posture. The **medulla oblongata** contains reflex centres which control a number of autonomic activities.

The **spinal cord** consists of a hollow central cylinder of **grey matter** composed primarily of neurone cell bodies. This is surrounded by **white matter** containing nerve fibres. Ascending tracts of fibres carry sensory information to the brain, while descending tracts conduct motor information from the brain. 31 pairs of segmentally arranged mixed **spinal nerves** emanate from the spinal cord. Sensory fibres enter the spinal cord via the **dorsal root** and motor fibres leave it via the **ventral root**. Somatic fibres supply the skin and voluntary muscles, visceral fibres supply the gut, and involuntary muscle and glands. Visceral motor fibres form the autonomic nervous system.

(b) The Autonomic Nervous System (ANS)

This is composed of *unmyelinated* nerve fibres which supply smooth muscle. The overall control of the ANS is maintained by the hypothalamus and medulla of the brain. The ANS is divided anatomically and functionally into two parts. The **sympathetic nervous system** possesses fibres that arise from paired segmental ganglia on either side of the spinal cord in the thoracic and lumbar regions. These ganglia connect with the cord via the **ramus communicans**. This system prepares the body for stressful situations. It accelerates heartbeat, dilates the iris, and diverts blood from the mesenteries to the brain and muscles.

The **parasympathetic nervous system** consists of the **vagus nerve** which arises in the medulla of the brain, and other nerves emerging from the sacral region of the spinal cord. This system regulates the basic activity of the body, e.g. heartbeat rate is kept normal.

Together, these two systems have opposite effects on the organs and glands they innervate. This enables the body to be maintained in a steady state (homeostasis) and for rapid and precise adjustments to be made in response to changing conditions.

(c) Simple Reflex Action

This is the basic functional unit of the nervous system. It may be **cranial** or **spinal**. Essentially, impulses received by a **receptor** are transmitted by **sensory neurones** to relay neurones in the grey matter of the spinal cord via the **dorsal root**. The **relay neurones** pass on the impulses to **motor neurones**. These leave the spinal cord via the **ventral root** and convey the impulses to the **effectors**. Individual reflex arcs are interconnected with others by multipolar relay neurones located in the white matter. Thus a response to a single stimulus could, and generally does, involve a large number of motor fibres.

Receptors exhibit the following common characteristics:

1. They respond to changes in the external or internal environment.
2. They transform the stimulus energy into an electrical response which then initiates a nerve impulse.
3. Different receptors produce the same kind of impulse whatever the stimulus.
4. Stimuli of varying intensity are translated into a varying frequency of impulses.
5. Repeated stimulation of a receptor leads to its reduced sensitivity.
6. Interpretation of impulses generated by the receptors is carried out by the brain.

9.3   Receptors

Exteroceptors are stimulated by external stimuli, **interoceptors** respond to internal stimuli, including changes in the homeostatic state, while **proprioceptors** are concerned with balance and the relative position and movements of the muscle and skeletal systems.

When a stimulus is received, the generator region of the receptor is depolarized, causing the production of a **generator potential (GP)**. The magnitude of the GP varies with the strength of the stimulus. When the GP exceeds the threshold level it gives rise to an action potential (AP). As the magnitude of the GP increases, more AP's per unit time are generated. Thus the *frequency* of nerve impulses is directly related to the strength of the stimulus and forms the means by which a graded response is made to varying stimulus intensity.

If a stimulus is continuously received by a receptor, eventually the GP falls below the threshold level and AP's cease to be propagated. This is called **adaptation**. It protects the organism by preventing it being overloaded with irrelevant information and means that organisms can ignore continuous background information while concentrating on more important stimuli.

9.4 Photoreception in the Eye

Rods are the main photoreceptors and are located in the **retina**. They function as low-light-intensity receptors and are not colour-sensitive. Light causes the conversion of **rhodopsin** into **retinene** and **scotopsin**. This evokes a GP which in turn causes the propagation of an AP. The light reaction is rapid. Resynthesis of rhodopsin involves energy which is provided by numerous mitochondria in the cell. Rods exhibit **convergence**, about 300 rods synapsing with one optic neurone. This reduces **visual acuity** but provides increased collective sensitivity under conditions of dim light through the process of **summation**.

Cones possess **iodopsin** which is only broken down under conditions of high light intensity. The closely packed cones in the **fovea** show no convergence, each cone being connected to its own optic neurone. Visual acuity is high. More cones are exposed to the focused image and the eye is able to see clearly objects which are very close together.

There are three functionally distinctive cones able to detect the colours blue, green and red. When the cones are stimulated to varying degrees, a particular colour is initially discriminated by the retina. However, the final colour perceived is determined by the brain. The absolute frequency of impulse discharge from the three types of cone determines colour intensity, while the relative frequency of impulse discharge determines the quality of the colour. Shortage or absence of a particular cone can lead to colour weakness or colour blindness.

Binocular vision occurs when the visual fields of both eyes overlap. This means that the fovea of both eyes are focused on the same object. **Stereoscopic vision** occurs when the two eyes simultaneously produce slightly different images. The visual cortex of the brain then resolves them as one image.

9.5 Hearing

Sound-waves cause vibrations of the **tympanic membrane** which are transmitted by the **ossicles** of the middle ear to the **oval window**. This vibrates correspondingly and causes displacement of fluid in the inner ear and movement of the **basilar membrane** of the **organ of Corti**. As the basilar membrane oscillates **receptor hair cells** attached to it are distorted mechanically. This causes depolarization and impulse propagation. The impulses are passed to the brain via the **auditory nerve**. The vibrations of the fluid in the inner ear are finally dissipated into the middle ear by vibrations of the **round window**. *Pitch* is determined by the frequency of the sound waves. Low-frequency sounds stimulate cells at the apex of the **cochlea**, high-frequency sounds stimulate those at the base of the cochlea. *Intensity* of sound depends upon the amplitude of the sound waves produced at source. **Tone** depends upon the number of different frequencies making up the sound.

9.6 Balance

This is controlled by the **vestibular apparatus** of the ear which consists of three **semicircular canals** placed at 90° to each other, the **sacculus** and the **utriculus**. At one end of each semicircular canal is a swollen **ampulla** containing the receptors. Hair-like projections from them are embedded in a gelatinous **cupula**. *Rotational* movements of the head cause displacement of the cupula in a direction opposite to the head movement. The receptor cells are distorted, produce generator potentials and fire impulses. **Maculae** in the sacculus and utriculus walls have hair-like receptor-cell extensions embedded in a gelatinous mass containing calcium carbonate granules called an **otoconium**. The utriculus and sacculus provide information about *vertical* and *lateral* movements of the head respectively and operate in a similar manner to the ampullae. Variable movements produce differences in the pattern of impulses fired by the receptors.

9.7 Chemical Communication

Hormones are secreted into the bloodstream by ductless **endocrine glands**. They exert their effects on target organs at a distance from the site of their production. Different hormones may interact with each other in a co-ordinated manner, e.g. the pituitary gland secretes thyroid stimulating hormone (TSH) which in turn stimulates secretion of thyroxine by the thyroid gland. Gradual increase in thyroxine concentration inhibits further secretion of TSH by the pituitary gland. However, when thyroxine concentration decreases, the pituitary gland is no longer inhibited and begins to secrete more TSH again.

This shows the endocrine system as an important agent of homeostasis. Secretion of hormones is controlled by the **negative-feedback** principle, though the whole system is under the ultimate control of the **hypothalamus**.

When a hormone reaches a target cell it may exert its effect at the level of the cell membrane, enzymes located in the cell membrane, cellular organelles or genes. A well-known reaction is that involving cyclic AMP. Hormones become attached to specific receptor sites on the target-cell membrane. This causes **adenyl cyclase** activity and the synthesis of cyclic AMP (C-AMP). C-AMP activates either the appropriate enzyme system or else genes which then evoke the response. This usually takes the form of altered metabolic activity.

Factors which control the release of hormones are:

1. The presence of specific substances in the blood, e.g. glucose and insulin concerned with blood sugar levels.
2. The presence of another hormone in the blood, e.g. TSH from the anterior lobe of the pituitary which stimulates secretion of thyroxine from the thyroid gland.
3. Stimulation by the autonomic nervous system, e.g. impulses stimulate adrenaline flow from the adrenal medulla of the adrenal glands.

9.8 Worked Examples

9.1

Explain the following terms relating to a nervous system:

(a) resting membrane potential The potential difference across the axon membrane produced by the maintenance of differential electrochemical gradients of K^+ and Na^+ ions in a resting axon.

(b) action potential The potential difference across the axon membrane produced by a sudden increase in the axon permeability to Na^+ ions which enter the axon. It produces a propagated depolarization.

(c) refractory period The period when an axon membrane cannot respond to further depolarization. This may be absolute or relative and is a limiting factor in determining speed of conduction.

(d) saltatory conduction in a myelinated axon At nodes of Ranvier, breaks in the myelin sheath occur. Here local circuits are set up and current flows. Action potentials 'jump' from node to node speeding up conduction.

(e) motor end-plate The region where the axon of a neurone runs in troughs in the muscle fibre. A large surface area of contact is established where transmitter substance is released, leading to production of an end-plate potential.

[10]
[AEB]

9.2

(a) How does the ionic balance within a resting nerve cell differ from that outside a nerve cell?

inside is negatively charged with high $[K^+]$, low $[Na^+]$, outside is positively charged with low $[K^+]$, high $[Na^+]$

[2]

(b) At a point on the axon of a mammalian nerve cell give the:

(i) nature of the resting potential across the membrane -60 mV [1]

(ii) factor initiating an impulse change in potential difference [1]

(iii) changes then occurring in the membrane sudden increase in Na^+

119

permeability followed by increase in K^+ permeability. Inside of
axon becomes positively charged and outside negatively charged.

.... [1]

(iv) ionic movements during the passage of the impulse . . . . Na^+ ions diffuse inwards and K^+ ions diffuse outwards. At a late stage the cation
pump reverses these movements.

.... [2]

(v) electrical changes during the impulse . . . . . . Inside of axon becomes . . .
positive for about 0.5 s.

.... [2]

(c) What changes occur in the nerve cell during recovery after an impulse? . . . .
Na^+ ions are pumped out of axon, K^+ ions are pumped into axon
by cation pump and the inside of the axon again
becomes positive.

.... [2]

[OLE]

9.3

(a) By means of labelled diagrams, show the structure of a neurone and of a synapse. [5, 5]

(b) Compare the mechanism of transmission of information along a neurone with that
across a synapse. [8]

[UCLES]

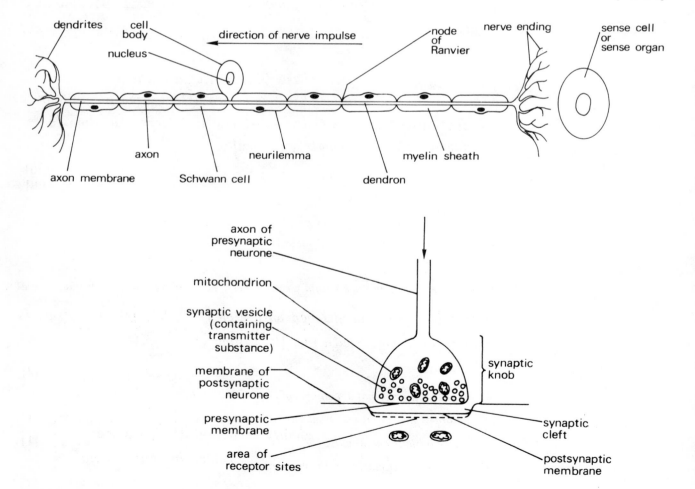

120

(b) Information passes along a neurone as a non-decremental frequency code of electrical impulses. At the synapse this information causes the release of chemical transmitter substances, such as acetylcholine or noradrenaline, which diffuse across the synaptic cleft and give rise to a propagated series of electrical impulses in the next neurone. The electrical impulses are called action potentials and they arise in the axon because of changes in the permeability of the axon membrane to sodium (Na^+) and potassium (K^+) ions. The release of transmitter substance by the presynaptic membrane also involves the presence of ions but in this case they are calcium (Ca^{2+}) ions. A cation pump in the axon membrane is necessary for the maintenance of electrical activity in the axon. Such a mechanism is not necessary at the synapse where the movement of transmitter substance is passive. The cation pump is necessary to restore the resting potential in the axon, whereas in the synapse the activity of released transmitter substance is destroyed by enzymes in the postsynaptic membrane, e.g. acetylcholinesterase.

The transmission of information along the axon is unidirectional by nature of the refractory period imposed by permeability changes. Transmission of information across the synapse is also unidirectional by nature of the methods of secretion and reception of transmitter substance across the synaptic cleft.

Finally, neurones, by nature of their ability to conduct all-or-nothing impulses, are excitatory. Synapses can be either excitatory or inhibitory, depending upon the nature of the transmitter substance.

9.4

(a) *State two ways in which the functions of the autonomic nervous system differ from those of the other parts of the nervous system.*
(i) not under conscious control of higher centres of brain
(ii) controls activities of the internal environment only

(b) *The roles of the sympathetic and parasympathetic systems within the autonomic nervous system are often antagonistic. Give one example of this type of antagonistic control and explain the precise roles of the sympathetic and parasympathetic systems in your example.*
Control of rate of heartbeat and amplitude of beat. Sympathetic system secretes noradrenaline which increases rate. Parasympathetic system secretes acetylcholine which decreases rate.

(c) *Name the transmitter substances which are produced by the autonomic nervous system.*
noradrenaline and acetylcholine

[7]
[AEB]

9.5

Figure 9.1 shows a lateral view of the human brain.

(a) (i) *Name the surface regions labelled A and B.*
A premotor area of cerebral cortex
B visual sensory area of cerebral cortex

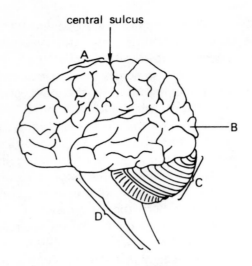

central sulcus

Fig. 9.1

(ii) *State one probable effect on human behaviour of a malfunctioning of*

region A .. *lack of control of complex co-ordinated movements* ..

region B .. *blindness*

(b) (i) *Name the part labelled C.*

.. *cerebellum* ..

(ii) *State the main function of part C.*

.. *integration of information for co-ordinated muscular activity* ..

(c) (i) *Name the part labelled D.*

.. *medulla oblongata* ..

(ii) *State one function of part D.*

.. *regulation of autonomic activities e.g. heart-rate* ..

[8]

[AEB]

9.6

(a) *Figure 9.2 shows a cross-section through the human eye.*

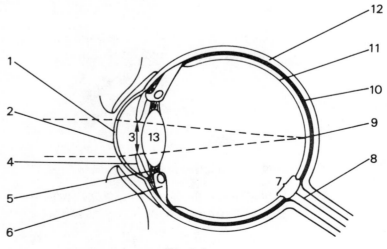

Fig. 9.2

Identify the features labelled 1 to 13. [2]

(b) Describe how the eye is able to accommodate to both near and distant objects. [2]

(c) Ophthalmologists often place a few drops of atropine (an antagonist of acetyl choline) in the eyes of their patients before performing a retinal examination.
    (i) What do you deduce is the effect of atropine on the pupil?
    (ii) What is the mechanism of this action? [4]

(d) In the human retina there are approximately 150 million sensory cells (7 million cones, the rest being rods). There are only 1 million fibres in an optic nerve.
    (i) What is the implication of these figures?
    (ii) With the aid of a simple labelled diagram of the human retina, explain the structural basis of visual acuity. (Visual acuity is a measure of the finest detail which the unaided eye can distinguish.) [7]

(e) (i) Give a brief account of the trichromatic theory of colour vision.
    (ii) Outline the experimental evidence for this theory. [5]

[NISEC]

(a)   1. cornea
      2. conjunctiva
      3. pupil
      4. iris
      5. suspensory ligament
      6. ciliary muscle
      7. blind spot
      8. optic nerve
      9. fovea centralis
    10. choroid layer
    11. retina
    12. sclera
    13. lens

(b) To focus on near objects the circular ciliary muscles contract, relieving the tension in the suspensory ligament, thereby allowing the natural elasticity of the lens to decrease its radius of curvature. This increases the refraction of light through the lens and brings a near image into focus on the retina. When the circular ciliary muscles relax the tension in the suspensory ligament increases, the radius of curvature of the lens increases, refraction is reduced and the parallel rays of light from distant objects are brought to focus on the retina.

(c) (i) If acetylcholine causes the pupils to constrict and atropine is an antagonist of acetylcholine, atropine must cause the pupils to dilate.

    (ii) Acetylcholine is released by the parasympathetic nervous system and stimulates the circular iris muscles to contract. Atropine causes the circular iris muscles to relax, thus dilating the pupils.

(d) (i) Several sense cells are connected to single sensory neurones, thus producing a high degree of sensitivity. In this example many rods must supply a single optic neurone. This is called convergence. Likewise, the combined effect of the simultaneous stimulation of several cells is cumulative and produces an effect in the optic neurones called summation.

    (ii) Visual acuity in humans is partly due to lateral inhibition but chiefly due to the anatomical arrangement of cones in the retina. 95 per cent of the 7 million cones in the eye are packed into the 1 mm diameter fovea in the centre of the retina. This close packing increases visual acuity. In addition, each cone is connected by a bipolar neurone to its own sensory neurone in the optic nerve as shown in Fig. 9.3.

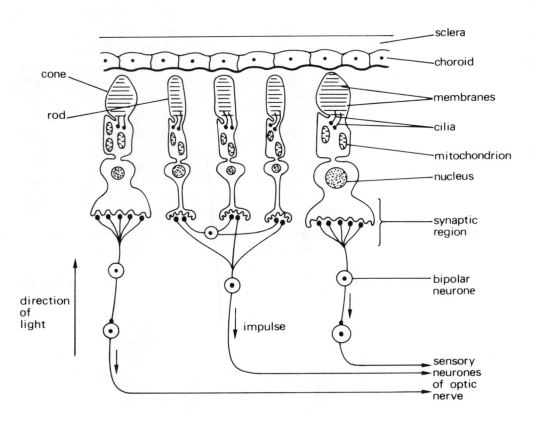

Labels in figure:
- sclera
- choroid
- cone
- rod
- membranes
- cilia
- mitochondrion
- nucleus
- synaptic region
- bipolar neurone
- direction of light
- impulse
- sensory neurones of optic nerve

Fig. 9.3

This absence of convergence and close packing produces high visual acuity.

(e) (i) There are three types of cones in the human retina, each possessing a different pigment, which absorbs light of a different wavelength. There are red, green and blue cones. Different colours and shades are produced by the degree of stimulation of each type of cone by light reflected from an object. Equal stimulation of all cones produces the colour sensation of white. While the initial discrimination of colour occurs in the retina, the final colour perceived involves integration within the brain.

(ii) Fine beams of light have been passed through sections of a retina and the wavelengths of light absorbed measured. Red, green and blue are always missing from the spectrum of light produced. This suggests that these corresponding wavelengths of light have been removed.

9.7

Figure 9.4 shows a section through the middle ear and inner ear of a mammal.

(a) Identify parts 1 to 4.

1 ......*cochlea*..

2 ......*ampulla*..

3 ......*stapes*..

4 ......*tympanum*..

(b) Give one function of each of the following:

part 1 ......*transduction of pressure waves into nerve impulses*............

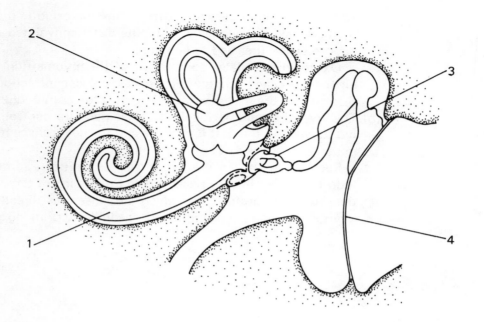

Fig. 9.4

part 2 ....`detect direction and rate of displacement of head`....

The middle ear ossicles ....`transmit and amplify vibrations of tympanum`....

(c) (i) Draw a labelled diagram to show the sensory receptors of part 2.

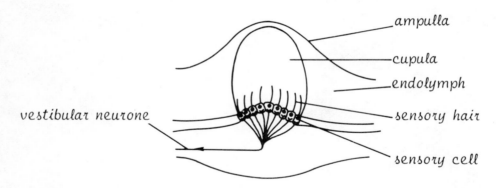

(ii) Explain how these receptors are stimulated.

`Movement of head is resisted by endolymph – this displaces`

`cupula – sensory hairs stimulate generator potentials in sense`

`cells.`

[9]

[AEB]

9.8

(a) For two named hormones synthesized by the hypothalamus and released from the pituitary, give their effects on their target organs.

(b) Describe two functions of the hypothalamus other than that of producing hormones.

[OLE]

(a) Antidiuretic hormone (ADH) maintains the osmotic pressure of body fluids by increasing the permeability of the distal convoluted tubule and collecting duct to water, urea and sodium ions.

Oxytocin stimulates contraction of the myometrium of the uterus during birth. Following birth, oxytocin release causes contraction of the myoepithelial tissue of the breast during lactation. This forces milk out of the nipples.

(b) The hypothalamus is involved in temperature control. Two regions of the hypothalamus, the heat loss centre and the heat gain centre, act as thermostats in the feedback control of body temperature.

Specific centres in the hypothalamus control essential functions such as food and water intake. Receptors in other parts of the body provide information to the hypothalamus, which also measures directly various metabolite levels in the blood which influence feeding and drinking activities.

10 Support and Movement

10.1 Skeletal Structures

Nearly all animals possess some form of supporting structure. This varies from small strengthening rods in protozoa to the complex vertebrate endoskeleton. Besides supporting the soft body-tissues skeletons may protect delicate internal structures and generally aid locomotion. Animals need to move from place to place

1. to capture food,
2. to avoid predators,
3. to seek a mate during reproductive activity and
4. for dispersal and colonization of new areas.

Such locomotion occurs through integrated activity between the animal's nervous, muscular and skeletal systems.

Three major types of skeleton exist. The earthworm possesses a **hydrostatic skeleton**. Here the circular and longitudinal body wall muscles contract against internal coelomic fluid. The muscles work antagonistically against each other, the net result being the production of peristaltic waves along the whole length of the worm, causing locomotion. Segmentation also enables the worm to move or change the shape of parts of its body only.

Arthropods possess an **exoskeleton** of **chitin**. At the hinge joints of these animals, where flexibility is necessary, the chitin is reduced to a thin membrane and antagonistic flexor and extensor muscles move adjacent segments of each joint. The exoskeleton is effective in support, protection and providing muscle attachments. However, it limits the size attained by the animals due to weight problems, and necessitates growth by ecdysis.

The **endoskeleton** in vertebrates is a living, internal part of the body which alters shape according to the changing needs of the growing organism. Individual bones are designed for muscle attachment, use as levers and the formation of **joints**. The endoskeleton consists of **cartilage** and/or bone. The three types of cartilage which exist all consist of a firm matrix secreted by **chondroblasts**. **Hyaline cartilage** forms the skeleton of the embryo but in the adult bony skeleton it is located at the ends of bones at articulating joints and at the ends of ribs. **White fibrous cartilage** is strong and has a degree of flexibility. It is a component of tendons and is present between vertebrae. **Yellow elastic cartilage** possesses strength and elasticity and is found in pharyngeal cartilages and the pinnae.

Bone is a tough, resistant calcified connective tissue. Its components are arranged to withstand both compression and stretching forces. Some parts of the bony skeleton manufacture granulocytes and erythrocytes. This skeleton also acts as a reserve of Ca^{2+} ions and phosphorus.

The mammalian skeleton consists of **axial** (skull, vertebral column, ribs and sternum) and **appendicular** (girdles and bones of limbs) components. The anterior portion of the axial skeleton is the cranium consisting of flat interlocking bones, which protect the brain and sense organs. The upper jaw is fused to the cranium while the lower jaw articulates with it via a hinge joint.

The **vertebral column** consists of a series of **vertebrae** held together by ligaments and separated from one another by intervertebral discs. There is some degree of movement between adjacent vertebrae and as a whole the vertebral column forms a flexible axis which supports the body. While each vertebra conforms to a general design there is a high degree of variability between them in different regions of the column. This reflects the different functions performed by each region. The vertebral column protects the spinal cord, and each vertebra has several bony projections used for muscle attachment. The thoracic vertebrae articulate with the ribs. When the intercostal muscles contract, causing movement of the ribs, breathing takes place. Together with the sternum, the ribs protect the heart, lungs and major blood-vessels.

The girdles of the appendicular skeleton connect the limbs to the axial skeleton. The two separate halves of the **pectoral girdle** are flexibly attached to the axial skeleton. This design provides for a wide range of movement for the forelimbs. It separates the limbs, thus contributing to a quadruped's support and stability and also functions as a shock absorber when the animal lands at the end of a jump.

The two halves of the **pelvic girdle** are fused together, and the girdle itself is fused to the sacrum of the vertebral column. This girdle helps to transmit the backward thrust of the hind limbs against the ground to the body. Mammalian/ vertebrate limbs are designed on the **pentadactyl limb** plan and exhibit the phenomenon of homology. They are structures with similar origin and general form but have been modified considerably to carry out numerous dissimilar functions.

The region where two bones meet is called a **joint**. Joints may be movable or immovable. At a moving **synovial** joint, when muscles contract, the bones move around the joint with respect to each other and are held in position by **ligaments**. The end of each articulating surface is covered with **hyaline cartilage**. This, together with **synovial fluid** secreted by the synovial membrane, lubricates and reduces friction as the bones move over each other. Synovial joints are classified according to the range of movement they provide.

10.2 Muscle Systems

Muscles convert chemical energy into kinetic or mechanical energy. They are composed of fibres, capable of contraction, extensibility and elasticity. Muscles are innervated and supplied with blood, the volume of which is regulated according to needs.

Cardiac muscle is myogenic, never fatigues and is innervated by the autonomic nervous system. **Smooth or involuntary muscle** is found in the walls of tubular organs and moves materials through them. It contracts and fatigues slowly and is innervated by the autonomic nervous system. **Skeletal** or **voluntary muscle** is attached to the bony skeleton and is primarily concerned with locomotion. It contracts and fatigues quickly and is innervated by the voluntary nervous system.

Skeletal muscle is attached to bone by relatively inextensible collagenous **tendons**. At one end, its **origin**, the muscle is attached to a part of the skeleton which does not move during muscle contraction. At its other end, the **insertion**, the muscle is attached to a skeletal component, further away from the main axis of the body, which is moved when the muscle contracts. Pairs of **antagonistic muscles** move a bone into one position and back again. Generally, groups of muscles called **synergists** operate in this way instead of just two muscles.

Each muscle is made up of many **muscle fibres**. Every fibre is divided into **sarcomeres**, each of which contains proteinaceous actin and myosin **myofibrils** bounded by a **sarcolemma**. The **sarcoplasm** contains numerous mitochondria and is permeated by a network of membranes forming the **sarcoplasmic reticulum**, and a system of transverse tubules, the **T-system**. At intervals these tubules pass between pairs of vesicles involved in Ca^{2+} ion movement. One tubule and a pair of vesicles constitute a **triad**.

Actin is composed of two molecules of **G-actin** (each bound to one ATP molecule) wound round each other to form **F-actin**. **Tropomyosin** which switches the contractile mechanism on or off, and **troponin-I** and **troponin-C** which collectively inhibit contraction in the absence of Ca^{2+} ions, also form part of the actin myofibril.

Myosin possesses two chemically active globular heads which, when activated, are the binding sites which attach to F-actin.

At rest there is a low concentration of Ca^{2+} ions in the sarcomere, and tropomyosin blocks the actin sites to which myosin can bind. When the sarcomere is stimulated the resulting depolarization causes release of Ca^{2+} ions from triad vesicles into the sarcoplasm. Ca^{2+} ions bind to troponin-C. This then interacts with troponin-I, causing the myosin sites to be unblocked and activated. Once activated, the myosin head moves out and binds to actin, forming an actomyosin cross-bridge. Hydrolytic breakdown of ATP accompanies **cross-bridge formation** and the energy released causes the myosin head to pull the actin filament towards the centre of the sarcomere. This leads to a shortening of sarcomere length and an overall contraction of the muscle. Cross-bridge formation and breakage is repeated many times and on each occasion a new bridge is formed it is between the myosin head and another actin subunit further along the myofibril. After stimulation an active cation pump returns the Ca^{2+} ions to the triad vesicles, reduction in the level of Ca^{2+} ions in the sarcoplasm occurs, and relaxation of the sarcomere begins.

In the presence of an adequate stimulus the fibre contracts maximally. No further increase in strength of the stimulus will produce a stronger contraction. This is called the 'all-or-nothing' response. A latent period of 0.05 s elapses prior to muscle contraction. Contraction lasts for 0.1 s and is followed by a 0.2 s period of relaxation. During this time an absolute refractory period is followed by a relative refractory period.

When another stimulus is applied while the muscle is still responding to the first stimulus, mechanical summation occurs whereby a second contraction of greater force is elicited. A rapid series of stimuli provokes a sustained contraction called tetanus. This contraction is stronger and/or higher than that developed in response to well-spaced-out, individual stimuli. Tetany ends when the muscle fatigues.

Groups of muscle fibres, called motor units, are innervated by a single nerve fibre. Motor units contain variable numbers of fibres, but, once they are stimulated, all fibres in the unit contract. The smoothly co-ordinated graded response of whole muscles is brought about by asynchronous contraction of different motor units in antagonistic muscles or the variable frequency of stimulation of the motor units.

When a muscle contracts the small quantity of ATP present is rapidly used up. Replenishment of ATP occurs when ADP is reconverted to ATP by phosphocreatine breakdown. Later, after contraction has ceased, phosphocreatine is reconstituted by ATP regenerated by energy from oxidation of fatty acids and glycogen.

If a muscle becomes very active the respiratory and blood systems are unable to supply sufficient oxygen for the muscle's needs. Consequently, pyruvic acid is converted to **lactic acid** by the addition of H^+ ions and the muscle builds up an **oxygen debt**. Removal of lactic acid occurs when activity slows down or ceases. One-fifth of the lactic acid is oxidized by oxygen, now readily available. The energy liberated by this reaction is used to reconvert the remaining lactic acid, first to glucose and then to glycogen which is stored in the liver. Some of the glucose is returned to the muscle where it is stored as muscle glycogen. The oxygen debt is repaid when all the lactic acid has been removed from the body.

Two types of fibre exist which collectively enable the organism to maintain posture, move parts of the body and move at variable speeds. **Red tonic** fibres generate slow, long, sustained isometric contractions valuable for good posture. **White twitch** fibres provide immediate fast muscle contraction which is of survival value to predators attempting to catch prey, and prey attempting to escape.

Muscle spindles within skeletal muscle detect and respond to the degree of stretching of the muscle. When a muscle is stretched, extension of the spindle causes a stretch reflex. This prompts the muscle to contract automatically in an attempt to regain its original length. Muscle tone, the state of partial contraction of muscles, is brought about by a continuous production of nerve impulses causing small numbers of fibres to contract. Muscle spindles are also used in the graded voluntary movements of the organism. As the muscle spindle is stretched, more impulses are discharged. This, in turn, determines the extent to which the muscle contracts.

Tendons possess stretch receptors which respond to high tension by inhibiting muscle contraction. Thus the muscle is protected from contracting too strongly and causing itself to be damaged.

10.3 Worked Examples

10.1

Figure 10.1 is a drawing of the forelimb skeleton and part of the shoulder girdle of a mammal (rabbit).

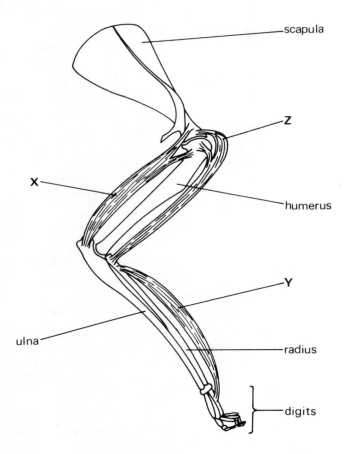

Fig. 10.1

Complete the diagram to show the following muscles and their attachments to the skeleton.

(i) A muscle which contracts in preparation for landing after a leap. Label this muscle X.

(ii) A muscle which will extend the digits. Label this muscle Y.

(iii) A muscle which will contract in response to a painful stimulus applied to the forefoot. Label this muscle Z.

[6]

[L]

10.2

Compare the skeletal systems of mammals, herbaceous plants and trees to show how:

(a) the organism is supported;

(b) growth occurs;

(c) the types of skeleton are adapted to the way of life of the organisms.   [OLE]

(a) Mammals are supported by a bony skeleton and muscles. The weight of the long bones of the limbs is reduced by being hollow and the bones are strengthened by the laying down of bone in concentric circles. The major bending stresses act on the periphery of the bones where strengthening trabeculae are most prominent. The long bones support the girdles which in turn support, and are supported by, the vertebral column. Vertebrae are held together by ligaments and each acts as a compression member of a strong but flexible girder supported by powerful back muscles attached to the transverse processes of the lumbar vertabrae.

Herbaceous plants, however, rely on cell turgor and specialized thickening materials in cells. Leaves and young stems rely for much of their support on the turgor pressure produced by vacuolar contents acting outwards on thickened cell-walls. In addition, bending strains in stems are countered by the thickening of collenchyma, sclerenchyma and lignified vessels. The peripheral arrangement of these supporting tissues also assists in resisting tension in stems. In roots, tension is developed as pulling strains which act centrally. The central position of the thickened tissues helps in resisting these strains.

Trees require a very rigid system of support provided by annular rings of thickened secondary vessels, tracheids and fibres as determined by the nature of the tree.

(b) Growth in the bones of mammals occurs by the invasion of cartilage by osteoblasts which synthesize bone collagen and lay down hydroxyapatite. In compact bone organic and inorganic materials are laid down in concentric rings or lamellae. This is called ossification. Growth in length of the bones also occurs by ossification which occurs at the ends of the bones. The outer region of the bone called the periosteum can also increase in thickness and increase the diameter of the bone.

In herbaceous plants the sites of growth are restricted to certain regions such as apical meristems and intercalary meristems. In these regions the undifferentiated meristem cells undergo mitosis to form initials. Cell expansion then occurs by vacuolation to give rise to larger cells prior to differentiation. This is called primary growth, in contrast to the secondary growth of trees. Here lateral meristems, composed of cells of the vascular cambium, divide to form secondary phloem and secondary xylem. The differential growth of spring and autumn xylem gives rise to annual rings. Increased growth, resulting in the lengthening of branches, occurs at apical meristems.

(c) In mammals one of the major functions of the skeleton is the protection of vital organs such as the brain by the cranium. The functions of movement and support are achieved by the co-ordinated action of bones and muscles. Movement of parts of the mammal occurs at joints. At some joints only limited movement can occur, as between vertebrae, but at other synovial joints a variety of movements are possible. Locomotion is achieved by long bones and joints forming levers. Muscle attachments are so situated as to increase the efficiency of the lever and support systems.

In herbaceous plants support is provided by thickened vascular tissue and turgor. Thickened tissue may occur in specific regions such as collenchyma in the corners of stems of *Lamium*. The petioles of leaves are thickened on the lower sides to provide sufficient support to expose leaves to the maximum available light. Flowers, too, must be prominently displayed in order to facilitate pollination and for the subsequent dispersal of fruits and seeds. These adaptations are necessary because of the sedentary nature of plant life.

Trees are subject to quite high stresses imposed by wind coming at different times and from many directions. The thickened xylem provides most of this support. The perennial pattern of growth produces increased girth to support the increasing height and branching. There is remarkable economy of structure and function shown by plants because the tissues described above have dual roles in transportation of solutes and in support.

10.3

Figure 10.2 is an electron micrograph of muscle tissue.

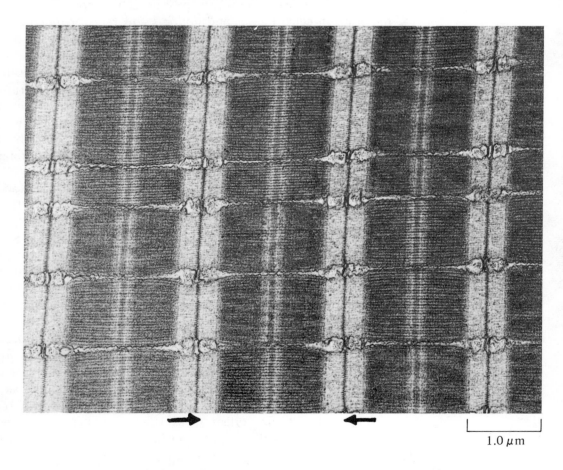

1.0 μm

Fig. 10.2 (Electron micrograph by courtesy of Dr J. M. Squire)

132

(a) What sort of muscle tissue is it and why is it so called?

striated muscle because of the striated appearance seen under the microscope.

(b) (i) What is the region between the two black lines delimited by the arrows called?

a sarcomere

(ii) Draw below a simplified model of this region and label the structures and bands shown.

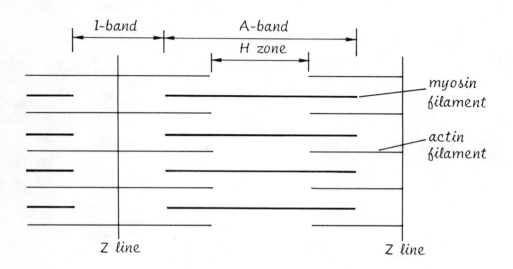

(iii) Outline how this model might work in allowing the muscle to be alternately relaxed and contracted.

During contraction actin filaments slide inwards over the myosin filaments. Cross bridges form. This is excitation-contraction coupling. Filaments slide apart during relaxation.

(iv) Is the muscle in the figure fully contracted, about half contracted or fully relaxed? State why.

About half contracted because the H zone is not very wide.

[7]
[UCLES]

10.4

Describe the sequence of events involved in the stimulation and contraction of a skeletal muscle fibre. (Start your answer at the point where the impulse reaches the end of the motor fibre.)
[10]
[L]

At the motor end-plate acetylcholine is released into the synaptic gap which separates the motor fibre from the sarcolemma of the muscle fibre. A reaction between acetylcholine and receptor sites on the sarcolemma increases the permeability of the latter to Na^+ ions and K^+ ions. This produces a wave of depolarization which spreads along the sarcolemma and down the T system to the sarcomere. As the impulse passes the triad vesicles it causes the release of Ca^{2+} ions into the sarcoplasm. The Ca^{2+} ions bind with troponin and prevent its inhibitory effect on the myosin binding sites. Thus activated, the myosin head moves out from its resting position and links to actin to form an actomyosin cross-bridge. In the presence of energy from ATP there is a change in the angle of the cross-bridge so that the myosin head pulls the actin filament over itself towards the centre of the sarcomere. This effect, amplified by the action of many more myofilaments, leads to a reduction in sarcomere length and the development of a tension within the muscle. This is called excitation–contraction coupling and generates the force necessary for the contraction of skeletal muscle.

11 Homeostasis

Negative feedback This opposes any detected tendency away from the optimal level and returns it to the optimal level.

Positive feedback This reinforces a particular tendency so that it moves further and further away from the level at which it is normally held.

Homeostasis The maintenance of the internal environment within certain narrow limits.

Ectotherm An organism that relies on heat derived from the environment to raise its body temperature.

Endotherm An organism that maintains a fairly constant temperature independent of the environmental temperature. It generates heat, which must be conserved, via a high metabolic rate.

Core temperature The temperature of the tissues below a level of 2.5 cm beneath the surface of the skin.

Basal metabolic rate (BMR) The amount of heat energy per unit mass released in a resting, fasting organism.

Glycogenesis The conversion of carbon residues arising from metabolism to glycogen.

Glycogenolysis The conversion of glycogen to glucose.

Organisms are able to maintain their internal environment in a steady state despite constant fluctuations in the external environment. This is called **homeostasis**. It provides cells with optimum conditions in which to function effectively and efficiently and provides organisms with varying degrees of independence of their environments.

Fluctuations either side of the **optimal level** (**reference** or **set point**) of a particular condition are detected by receptors which stimulate control centres in the brain to counteract, correct and return that condition towards its optimal level. This is known as **negative feedback** and leads to the stability of systems. **Positive feedback** is rare in biological systems and generally leads to instability.

11.1 Control of Respiratory Gases in the Blood

Normally there is a **partial pressure** (p.p.) of 100 mmHg of oxygen and 40 mmHg of carbon dioxide in the blood. These levels are maintained by negative feedback through operation of **stretch receptors** in the lungs and trachea, and **chemoreceptors** in the **carotid bodies** of the external carotid arteries and in the medulla.

Ventilation is controlled by **respiratory centres** in the medulla and pons. *Depth* and *rate* of ventilation are determined by the responses of chemoreceptors and in turn the respiratory centres, to changes in the p.p. of carbon dioxide in particular. Only when the p.p. of oxygen drops to a very low level do the chemoreceptors respond to the p.p. of oxygen.

The level of carbon dioxide in the blood also affects the rate of circulation. An increase in blood carbon dioxide leads to more impulses being transmitted from the **cardiovascular centre** of the **medulla** to arterioles which respond by constricting, thereby raising blood-pressure and speed of blood-flow. Conversely, an increase in carbon dioxide levels in tissues and organs (e.g. muscles) causes their arterioles to dilate.

Blood-pressure is under the control of **stretch receptors** in the **carotid sinus**. If blood-pressure rises, the receptors are stretched. They increase impulse transmission to the cardiovascular centre, which in turn sends impulses via efferent nerves to the heart **sino-atrial node**. This slows down heart-rate and therefore reduces blood-pressure. Peripheral arterioles vasodilate, further reducing blood-pressure. If blood-pressure does fall below normal limits the opposite actions occur in order to raise it.

11.2   Temperature Regulation

The body temperature of an organism affects the rate of movement of atoms, ions and molecules and the rate of enzyme activity.

Heat energy is transferred along thermal gradients by **conduction**, **convection** and **radiation** between organisms and their environments. It may also be lost during evaporation of water.

Ectotherms cannot maintain their body temperatures within narrow limits and their activity is determined by the prevailing temperature of their surroundings. Therefore they must live within a narrow temperature zone of tolerance and must rely on gaining heat from their environment to raise their own body temperature. Ectotherms produce little internal heat of their own, have little insulation capacity and lack mechanisms for conserving heat.

Aquatic ectotherms have a body temperature equal to that of the water in which they live. This temperature varies only very slightly, whereas terrestrial ectotherms are subjected to much greater fluctuations. Consequently, terrestrial ectotherms exhibit behavioural activities in order to gain or lose heat and therefore achieve some degree of temperature control.

Endotherms (birds and mammals) maintain their body temperatures within narrow limits by physiological means. This provides optimum conditions for high metabolic activity and makes them relatively independent of the external temperature. Endotherms generate a lot of internal heat and are endowed with good insulation. They also possess a variety of mechanism which control heat gain and loss. These are under the control of the hypothalamus.

Surface temperature fluctuations are detected by free nerve endings in the skin, while changes in the core temperature of the body are detected by thermoreceptors in the hypothalamus, medulla and spinal cord.

When external changes are perceived the **hypothalamus** receives impulses from the **peripheral receptors**. It then sets up rapid appropriate corrective mechanisms that enable the body to maintain a constant core temperature. This is usually achieved by controlling the amount of conduction, convection and radiation, vasodilation and vasoconstriction and by shunting more or less blood to the skin as necessary.

(a) Mechanisms Occurring Under More Extreme Conditions

When conditions become more extreme the following mechanisms take place.

(i) *Overcooling*

1. **Insulation** is provided by a layer of still air between the hair or feathers. Air is a poor conductor of heat and therefore reduces heat loss. When the hairs or feathers are raised the insulating layer is increased and helps to reduce further heat loss. Aquatic endotherms already possess a thick layer of fat/blubber which provides effective insulation.
2. **Reduction of blood-flow** between skin and body core. Vasoconstriction of dermal arterioles is under the control of the vasomotor centre in the brain which receives impulses from the temperature regulation centre in the hypothalamus. This mechanism reduces heat loss by convection, conduction and radiation as well as sweat secretion and consequent heat loss by evaporation. Arterio-venous shunt systems also operate to divert blood from the skin capillary beds.
3. **Shivering** is a muscular activity designed to generate heat by increasing metabolic activity and therefore raising body temperature. Increased production of adrenaline and thyroxine also stimulates increased metabolic activity.

(ii) *Overheating*

1. **Reduced air-layer** insulation and subcutaneous fat layer.
2. **Increased blood-flow** between body core and skin caused by vasodilation of dermal arterioles. More blood near the skin surface increases heat loss via conduction, convection and radiation. Sweat production is increased and therefore more heat is lost by evaporation. This begins when the core temperature rises above 36.7 °C.

(b) Adaptations of Animals to Life at Low Temperatures

Excessive heat loss is prevented by **counter-current heat-exchange** systems in the appendages. This enables the colder blood returning to the body from the appendages to be warmed by outgoing blood. Low temperatures stimulate hibernation in many warm-blooded animals. Having laid down fat stores, their core temperature drops dramatically, and this is followed by a corresponding reduction in metabolism. Temperature regulation now operates at a lowered set point. Under conditions of food shortage this conserves energy. Hibernation comes to an end during a process of self-warming.

Ectotherms become torpid at reduced temperatures and recover when the external temperature warms them up again.

(c) Adaptations of Animals to Life at High Temperatures

Some animals are able to live satisfactorily over a wide range of temperatures. For example, the body temperature of the camel is very low (34 °C) during the night, but rises to 40 °C during the day. The camel is also very tolerant to dehydration and does not sweat.

Many fresh-water fish and amphibia burrow in the mud when their habitats periodically dry up. When water is present they break this condition. Invertebrates may produce spores or similar resistant resting stages in unfavourable conditions.

(d) Adaptations of Plants to Low Temperatures

Temperate deciduous woody perennials lose their leaves in autumn. This prevents water loss during winter when soil water is relatively unavailable. The thin, needle-like leaves of conifers reduce their surface area and therefore the area on which snow can accumulate. This also reduces wind and snow damage. Buds are protected by scale leaves, and their development in unfavourable conditions is inhibited by growth regulators. Resistant seeds and organs of perennation ensure survival and continuation of many flowering plant species. Substantial numbers of plants, however, require short exposure to low temperatures (e.g. 0 to 4 °C) to break dormancy and initiate the process of **vernalization**.

(e) Adaptations of Plants to High Temperatures

The process of transpiration reduces the risk of the plant overheating. If the temperature rises too high there may be a dramatic increase in transpiration rate as a consequence of stomata opening to their fullest extent (desert plants). The temperature is subsequently reduced owing to loss of heat energy since the evaporation of water requires the uptake of the latent heat of vaporization. Wilting reduces the surface area of a leaf exposed directly to the sun and reduces heat absorption and loss of water by transpiration. Plant leaves are generally thin with a large surface area. These features and shiny cuticle reflect much of the incident light. **Xerophytes** show morphological and physiological adaptations which enable them to survive.

11.3 The Liver

This organ controls many of the metabolic activities necessary for regulating and maintaining the constant composition of the blood. The liver contains large numbers of **hepatocytes, blood spaces** and **bile canaliculi**. It contains a large volume of blood and has a large surface area whereby exchange and control of materials between cells and blood can take place. Some of the homeostatic functions of the liver are summarized below.

(a) Carbohydrate Metabolism

The brain must receive a continuous supply of blood glucose at a level of 90 mg glucose per 100 cm^3 of blood. When the level falls below 60 mg per 100 cm^3 blood, **glycogenolysis** occurs. This takes place under the influence of the hormone **glucagon** secreted by the **alpha cells** of the **islets of Langerhans** of the pancreas, and stimulates the conversion of glycogen to glucose. **Adrenalin** can also cause glycogenolysis under conditions of stress.

If there is a high blood-sugar level **glycogenesis** occurs. The **beta cells** of the **islets of Langerhans** secrete **insulin** into the bloodstream which stimulates liver cells to convert glucose to glycogen and fat.

(b) Protein Metabolism

Surplus amino acids cannot be stored in the body. Therefore **deamination** occurs. This involves the removal of the $-NH_2$ group from the acid and the simultaneous oxidation of the rest of the molecule to form carbohydrate. The carbohydrate can then be used in respiration. **Keto amino acids** may be respired directly without having to be converted into carbohydrate. The $-NH_2$ group is converted first to **ammonia** and then to **urea** via the **ornithine cycle**. The urea is eventually excreted from the body dissolved in water, as **urine**. Insects and birds secrete **uric acid** rather than urine.

(c) Plasma Protein Production

Plasma albumin exerts a colloidal osmotic pressure which opposes the hydrostatic pressure developed in the blood-vessels. This helps to maintain the balance of fluids inside and outside the blood-vessels.

Plasma globulins are involved in the immune response while other plasma proteins form part of the blood-clotting mechanism.

(d) Detoxification

Kuppffer cells remove pathogens from the bloodstream. Toxins are broken down in the hepatocytes by oxidation, reduction, methylation reactions or by combination with another organic or inorganic molecule. The removal of toxins helps maintain the constant composition of blood.

Note: It is now thought erroneous to consider the liver as a major region of heat production for the body, as many of its coupled chemical reactions are endothermic. Under normal circumstances the liver is 1 to 2 °C warmer than the rest of the body core.

11.4 Worked Examples

11.1

(a) What is meant by homeostasis?

> Homeostasis is maintenance of the internal environment within certain narrow limits. [1]

(b) Explain how the following adaptations might assist in homeostasis:

(i) an elongated loop of Henlé in a desert mammal

> The concentration of urine increases with the length of the loop of Henlé. Water is vital to desert mammals and is conserved by reabsorption by the long ascending limb.

(ii) the thick fur pelt in an arctic mammal

> Thick fur traps a layer of air. Air is a poor conductor of heat and effectively insulates the body against heat loss in cold conditions.

(iii) the subcutaneous fat in a marine mammal

> Fat too is a poor conductor of heat and if present as a

..... thick layer it will prevent heat loss to the water. ....

....

[6]

(c) State three major processes by which water may be lost from a mammal and in each case give the reason for this loss.

(i) Gaseous exchange and expiration. Alveoli are permeable to gases and water.

(ii) Sweating. Water is a major component of sweat. Evaporation of the water cools the body.

(iii) Excretion. The removal of excess salts and urea from the body occurs in solution. Water is the solvent.

[6]

(d) Giving two examples in each case, describe how organisms (other than mammals) adapt to (i) daily and (ii) seasonal changes in temperature.

(i) daily

first example Crocodiles spend the night in water where it is warm. They come onto land during the day to raise their body temperature but return to water if too hot.

second example The desert locust aligns itself at right angles to the sun in order to absorb heat. If it becomes too hot it aligns itself parallel to the sun's rays.

(ii) seasonal

first example Lungfish burrow into the mud of rivers during hot dry periods. This is called aestivation and enables them to reduce their metabolic rate and survive.

second example Frogs and newts hibernate during cold weather. They go into a state of torpor as their metabolic activities slow down.

[8]

11.2

The scheme given in Fig. 11.1 for the regulation of glucose levels in the mammalian body illustrates two important principles of homeostasis.

(a) What two important principles of homeostasis are illustrated in the scheme?

(i) The system is self-regulating, i.e. changes in glucose level set in motion events which will return the glucose level to the normal level.

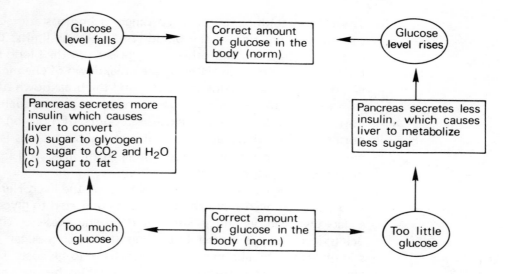

Fig. 11.1

(ii) .... *The system operates by negative feedback.* ....

(b) What type of substance (in terms of its function) is insulin? Give **three** *reasons for your answer.*

.... *hormone* ....

(i) .... *produced by an endocrine gland (pancreas)* ....

(ii) .... *transported in bloodstream to target tissues* ....

(iii) .... *exerts effect at point remote from origin* ....

[4]

11.3

Describe the role of the mammalian liver in
(a) protein metabolism,
(b) detoxification,
(c) carbohydrate metabolism.   [8, 6, 6]
[L]

(a) Amino acids cannot be stored in the body. Excess amino acids are deaminated in the liver by the enzymic removal of the amino group ($-NH_2$). The remainder of the molecule is then oxidized to form a carbohydrate which is utilized in respiration. The amino group is immediately reduced to form ammonia. Due to its high toxicity in aqueous solution ammonia is metabolized within a series of biochemical reactions known as the ornithine cycle.

In these reactions ammonia and carbon dioxide combine in a series of reactions to form urea which passes into the blood and is transported to the kidneys.

A number of amino acids which are deficient from the diet are synthesized by the enzymic transfer of the amino group from one amino acid to a keto acid. This is called transamination.

Finally, a number of plasma proteins such as albumins and globulins and blood-clotting factors such as prothrombin and fibrinogen are manufactured in the liver.

141

(b) The term detoxification covers a range of homeostatic activities which maintain the composition of the blood within certain limits. This may include the removal of bacteria and other pathogens from the blood by Kupffer cells but is usually thought of in terms of the breakdown of chemical substances such as alcohol to carbon dioxide and water and the breakdown of hydrogen peroxide to water and oxygen by catalase. Foreign substances such as drugs are broken down by the liver, as are excess amounts of hormones, in particular sex hormones. The removal of certain metabolites such as lactic acid and heavy metal salts from the blood also occurs in the liver and may be considered to be part of the process of detoxification.

(c) All hexose sugars from the gut pass through the liver. Here excess amounts of sugars are removed from the blood and converted to glycogen. This process is called glycogenesis and is stimulated by the presence of insulin. Glycogen, being insoluble, is stored but when the plasma sugar level falls glucagon stimulates the breakdown of glycogen back to glucose. This process is called glycogenolysis. In extreme cases of undernourishment both fats and proteins can be converted into sugars. This is called gluconeogenesis. In cases of overeating excess carbohydrates are converted into fats and stored in the body.

11.4

(a) How do animals regulate the composition of the blood? [12]

(b) Discuss the extent to which plants carry out homeostasis. [3]

[UCLES]

(a) Blood and a mechanism for its circulation are found in annelids, molluscs, arthropods and vertebrates. In all cases there are specialized tissues and organs for regulating the composition of blood, usually by the removal of substances either in excess or toxic. In annelids, excess ammonia and urea are actively pumped out of blood capillaries into the narrow tube of the nephridium. Arthropods have Malpighian tubules attached to the junction of the midgut and hindgut. Substances in excess in the haemolymph are removed into the tubules and then pass out of the body along with egested food. Marine birds, such as the penguin and albatross, ingest large quantities of salts along with their food and these need to be removed rapidly in order to prevent adverse effects upon the tissues. The excess salts are removed from the blood by specialized salt-secreting cells in the nasal glands. These glands secrete a sodium chloride solution which is four times stronger than the body fluids.

Mammals regulate the composition of respiratory gases, metabolic wastes, water and pH of the blood within narrow limits.

The levels of oxygen and carbon dioxide in the blood are regulated by adjusting the rate and depth of ventilation. Stretch receptors in the walls of the trachea and lungs and chemoreceptors in the walls of the aorta, the carotid bodies and the medulla regulate the ventilation rate by negative feedback. Variations in carbon dioxide levels stimulate the activity of these receptors. Information from these receptors passes to the respiratory centres in the pons and the medulla oblongata from where the rate and depth of ventilation are controlled. In addition, the partial pressure of carbon dioxide in the blood influences the affinity of haemoglobin for oxygen and thereby affects the rates of uptake and release of oxygen from haemoglobin.

The control of metabolites such as glucose, proteins and salts involves the integrated secretion of hormones mediated by receptor cells and the nervous system. In the case of glucose, for example, a rise in glucose level is detected and the cells of the islets of Langerhans release insulin in the blood. This influences muscle and liver metabolism and leads to a fall in blood glucose level mainly by the formation of glycogen. A fall in glucose level is detected by the islets, the adrenal glands, the medulla and the hypothalamus. Several hormones are released including glucagon and this causes the breakdown of glycogen and protein to produce glucose.

The control of plasma osmotic pressure involves osmoreceptors in the hypothalamus which stimulate anti-diuretic hormone (ADH) secretion. The permeabilities of the distal convoluted tubule and collecting duct can be increased or decreased by ADH, depending on whether the osmotic pressure increases or decreases. This, in turn, influences the volume and concentration of the urine produced. Aldosterone, too, influences water reabsorption by the kidney by stimulating the uptake of sodium ions from the filtrate into the plasma at times when water conservation is critical.

(b) It is difficult to compare reliably the extent to which plants and animals carry out homeostasis. Animal tissues tend to function within fairly narrow tolerance ranges whereas in plants the range of tolerance is much greater. Plant tissues are often much hardier and withstand greater physiological adversity than animals. For example, plant cells can lose a far higher proportion of water than animal cells and recover afterwards. One of the major problems encountered by plants is extremes of temperature. On hot, sunny days many plants show photosynthetic slump. This is a temporary reduction of metabolic activity resulting from changes in enzyme structure or the closure of stomata. The main reason for this may be water loss, although wilting is often seen in well-watered plants on hot days. The rate of water loss from the plant by transpiration is temperature-dependent and the loss of latent heat in this way prevents the plant from overheating. Ecological and anatomical adaptations of plants are outside the scope of homeostatic control in the strict sense of the term homeostasis and have been omitted from this discussion.

12 Excretion and Osmoregulation

Excretion   The elimination of the waste products of metabolism from an organism.

Secretion   The passage of *useful* intracellular molecules into the extra-cellular environment. (This term should not be confused with excretion, which deals with *waste* substances.)

Egestion   The elimination of waste substances (mainly undigested food), which have never been involved in cellular metabolic activities.

Metabolism   The chemical and physical reactions occurring in living organisms.

Osmosis   The movement of water molecules from a region of *their* high concentration to a region of *their* low concentration through a differentially permeable membrane.

Hypotonic   The state which exists within a solution which contains *more* water molecules than another solution from which it is separated by a differentially permeable membrane (i.e. a solution with a *higher* water potential than another solution).

Hypertonic   The state which exists within a solution which contains *less* water molecules than another solution from which it is separated by a differentially permeable membrane (i.e. a solution with a *lower* water potential than another solution).

Isotonic   The state which exists between two solutions separated by a differentially permeable membrane which have the same water potential (see Chapter 7).

Ultra-filtration   The process by which water and solute molecules separate from a solution, under pressure, according to their differential abilities to pass through a membrane.

Selective reabsorption   The selective passage of solute molecules, ions and water through a membrane.

12.1   Significance of Excretion and Osmoregulation

1. The removal of metabolic waste-products is necessary to ensure that chemical reactions proceed in their desired directions. For example, in the chemical reaction

$$A + B \rightleftharpoons C + D$$

the removal of C or D is necessary to ensure that the reaction proceeds from left to right.

2. The majority of waste-products are toxic and would damage cells if allowed to accumulate.
3. The ionic balance of the body tissue fluids must be maintained within narrow limits. This is achieved by the selective uptake or elimination of ions. This may affect also the water content of the body fluids.
4. The regulation of the water content of body fluids is one of the major physiological problems faced by organisms. The mechanisms of obtaining water, preventing water loss and eliminating water are diverse. They involve structural, functional and behavioural adaptations.

For the last two reasons it can be seen that excretion and osmoregulation are associated processes.

12.2   Excretory Products

The major excretory products of animals and plants are:

1. **Oxygen** produced by photosynthesis. This is used by plants and animals for aerobic respiration.
2. **Carbon dioxide** produced by aerobic and anaerobic respiration and by the breakdown of urea. This is a reactant in photosynthesis.
3. **Water** produced by aerobic respiration and by condensation reactions. It is the solvent for all metabolic activities and a reactant in photosynthesis.
4. **Tannins** and **other organic acids** produced in plants by metabolic activities.
5. **Bile salts and bile pigments** produced in the liver by the breakdown of lipids and haem respectively. The former are used in the emulsification of fats.
6. **Nitrogenous products** produced by the breakdown of proteins and nucleic acids. The immediate waste-product is **ammonia**. This is a very soluble, low-relative-molecule-mass molecule, NH_3, which is extremely toxic and cannot be stored. It must undergo immediate chemical change or be excreted.

   Urea is produced in the liver in a cyclical reaction, the **ornithine cycle**, from carbon dioxide, and ammonia produced by the **deamination** of amino acids. It is less toxic and less soluble than ammonia and is the major excretory product of elasmobranchs, some teleost fish, adult amphibia and mammals.

   Uric acid is insoluble and relatively non-toxic. It can be stored in cells without producing adverse toxic or osmoregulatory effects. Very little water is lost by organisms during the excretion of uric acid and it is the major excretory product of many terrestrial organisms such as insects, lizards, snakes and birds.

   Trimethylamine oxide is excreted in small amounts by a number of fish and gives fish their characteristic smell.

12.3   Mammalian Kidney

The kidney is the major excretory and osmoregulatory organ of mammals. It carries out the following functions:

1. removal of metabolic waste-products and 'foreign' molecules;
2. regulation of the chemical composition, water content, volume and pH of body fluids.

These functions ensure that the composition of tissue fluids is kept within narrow limits (an example of homeostasis). This enables cells to function efficiently and effectively at all times.

(a) Gross structure

Humans have a pair of kidneys situated at the back of the abdominal cavity. The **ureters** transfer **urine** formed in the kidneys to the **urinary bladder** where it is stored pending release from the body via the **urethra**.

Each kidney has the structure shown in Fig. 12.1.

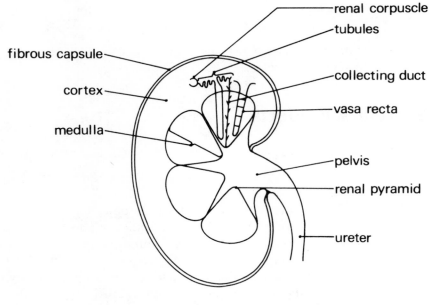

Fig. 12.1

Blood is brought to the kidneys from the aorta by the **renal arteries**. The **renal veins** return blood to the **inferior vena cava**.

(b) Nephron: Structure and Function

Each kidney contains approximately one million nephrons. The mean length of the nephron is 3 cm. Thus the kidney has an enormous surface area for the exchange of materials.

There are *six* main regions to each nephron. Each has a specific anatomical structure and physiological function. The overall function is the regulation of the composition of the blood.

(i) *Renal Corpuscle*

The double-walled **Bowman's capsule** encloses a knot of capillaries, the **glomerulus**. These capillaries originate from a single **afferent arteriole** and are drained by a single **efferent arteriole**. **Ultrafiltration** occurs here owing to the increased hydrostatic pressure which builds up in the capillaries. All substances with a relative molecular mass less than 68,000 leave the capillaries and form the fluid called **glomerular filtrate**. These substances filter from the blood through openings called **fenestrations** in the capillary walls. The **basement membrane** completely envelopes each capillary and forms the only structure separating the blood from the glomerular filtrate. The inner layer of Bowman's capsule is composed of cells called **podocytes**. These have extensions called foot processes which support the basement membrane and capillary. Between the foot processes are **slit pores** which facilitate the process of filtration.

About 125 cm³ of filtrate is produced per minute and this has a chemical composition similar to blood plasma. It contains glucose, amino acids, vitamins, hormones, urea, uric acid, ions and water. Albumins, globulins, white and red blood corpuscles and platelets are unable to pass out of the capillaries.

Of the 125 cm³ filtrate produced per minute, 124 cm³ are reabsorbed, leaving 1 cm³ which will eventually become urine. This reabsorption of filtrate occurs throughout the rest of the nephron, but principally in the proximal convoluted tubule.

(ii) *Proximal Convoluted Tubule*

About 80 per cent of the filtrate is reabsorbed into the blood in this region. The single layer of epithelial cells of the tubule has a **brush border** made up of **microvilli** on its inner surface. This increases the surface area for reabsorption. Substances diffuse through the membrane into the cells and are then pumped by active transport carrier mechanisms into the blood capillaries surrounding the tubule. Numerous mitochondria close to the cell membranes produce the energy required for this **selective reabsorption**.

(iii) and (iv) *Descending and Ascending Limbs of Loop of Henlé*

Desert-dwelling animals such as the kangaroo rat have much longer loops than related animals such as the beaver, which is semi-aquatic. The limbs of the loop of Henlé are concerned with the reabsorption of water and the production of urine. The loop of Henlé, the collecting duct and the capillaries of the vasa recta which surround these two structures create and maintain an increasing osmotic gradient in the medulla from cortex to pelvis. This gradient is produced by the presence of increasing concentrations of sodium and chloride ions and urea deep within the medulla. As filtrate flows through the loop of Henlé and the collecting duct, water is removed from them into the medulla by osmosis. From here the water is taken up into the capillaries of the vasa recta, again by osmosis. The maintenance of the osmotic gradient depends upon the **differential permeabilities** of the limbs of the loop which collectively function as a **countercurrent multiplier**. The permeabilities are shown in Fig. 12.2.

The tonicities of filtrate in the regions of the loop are also shown in Fig. 12.2. The **vasa recta** act as a **countercurrent exchanger** which enables the osmotic concentration of plasma *leaving* the kidney to remain at a steady state irrespective of the osmotic concentration of plasma *entering* the kidney.

(v) *Distal Convoluted Tubule*

The cells of this tubule have brush borders and numerous mitochondria and exercise a fine control of salt, water and pH balance of blood.

A decrease in blood volume causes the release of a hormone, **aldosterone** from the adrenal cortex. This stimulates the active uptake of sodium ions from the filtrate into the plasma of the capillaries surrounding the tubule. This uptake causes the reabsorption of an osmotically equivalent amount of water, thus restoring the blood volume.

The active excretion of either hydrogen ions or hydrogencarbonate ions, as appropriate, regulates the pH of the blood. The removal of hydrogen ions will increase the blood pH and the removal of hydrogencarbonate ions will decrease the blood pH. These changes are reflected in changes in pH of the urine.

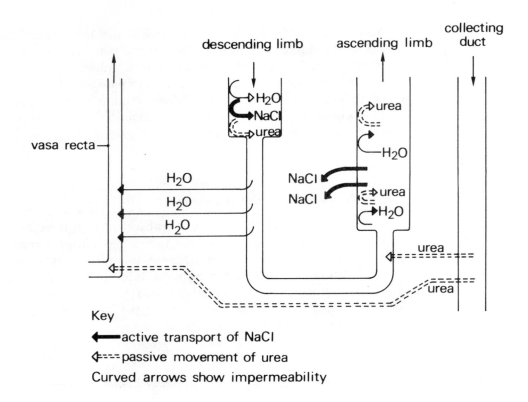

Fig. 12.2

(vi) *Collecting Duct*

The permeability of the walls of the duct to water and urea is controlled by **anti-diuretic hormone (ADH)**. ADH can also influence the permeability of the distal convoluted tubule. If the osmotic pressure of the blood rises, this is detected by **osmoreceptors** in the hypothalamus which cause the release of ADH from the posterior pituitary gland. ADH increases the permeabilities of the distal convoluted tubule and collecting duct to water which is withdrawn from the urine into the medulla and then absorbed into the blood. When this happens a reduced volume of **hypertonic** urine is excreted. Under normal conditions of water uptake in the diet no ADH is released, the tubule and duct remain impermeable and a relatively large volume of **hypotonic** urine is excreted.

12.4   Other Excretory Mechanisms

Ammonia and carbon dioxide diffuse out of protozoa and coelenterates through the entire surface. Freshwater protozoans are hypertonic to their environment and have specialized osmoregulatory organelles called **contractile vacuoles** which expel water which has entered by osmosis. In *Amoeba*, contractile vacuoles can appear anywhere in the cytoplasm but in *Paramecium* there are two vacuoles which have fixed positions.

Platyhelminths have a joint excretory–osmoregulatory system made up of tubules called **protonephridia** which end in enlarged hollow cells containing cilia. These cells are called **flame cells**. The movement of the cilia propels fluid along the tubules to their external openings. It is thought that waste enters the tubules by active transport.

Annelids, too, have a joint excretory–osmoregulatory mechanism composed of **nephridia**. Each nephridium is composed of a ciliated and muscular tubule leading from a ciliated funnel called a **nephrostome** to a bladder where waste is stored prior to release through a **nephridiopore**. The waste fluid is formed by ultra-filtration, selective reabsorption and secretion.

Terrestrial arthropods have specialized excretory organs called **Malpighian tubules** which produce and excrete the almost insoluble waste substance **uric acid**. The tubules are blind-ending and lie in the intercellular spaces of the abdomen where they are bathed in **haemolymph**. Many aquatic crustaceans excrete ammonia through organs called **antennal glands**.

All vertebrates have a form of kidney similar to that of mammals. In fish, the volume and concentration of the urine depends upon the osmotic pressure of the environment. Fresh-water teleosts excrete a large volume of dilute urine, while marine teleosts excrete a small volume of isotonic urine. Elasmobranchs increase their internal osmotic pressure so that it is isotonic with the seawater by retaining urea in the tissues.

Amphibia excrete urea but reptiles and birds excrete uric acid. The latter groups of organisms also produce a **cleidoic egg** during development. Uric acid is stored within the egg in a sac-like structure, the **allantois**.

12.5 Excretion and Osmoregulation in Plants

Plants synthesize all their organic requirements according to demand. There is never an excess of nitrogenous substances that require immediate excretion. The oxygen, carbon dioxide and water produced as waste metabolic substances are re-used in other metabolic processes. Excess oxygen produced during photosynthesis is however excreted. Excess salts and organic acids, e.g. oxalates and pectates, are stored as harmless insoluble products. Other acids, e.g. tannic and nicotinic acids, accumulate in leaves and are shed during leaf abscission.

Plants have fewer osmoregulatory problems than animals. Freshwater aquatic plants are called **hydrophytes**. They take in water by osmosis until their turgor pressure prevents further water uptake. Plants inhabiting areas of high salinity are called **halophytes**. Their root systems are able to tolerate high salinities. Many species have extensive roots which are able to store water when it is freely available. Other species can regulate their salt content by excreting salt from glands on the leaf.

Mesophytes occupy habitats with adequate water supplies. Their major problem, being exposed to air, is water loss. The presence of cuticle, protected stomata, variable leaf shape, abscission and ecological distribution according to tolerance to dehydration, aid survival.

Xerophytes survive long periods of drought. Some species survive in the seed or spore stage and germinate, grow, flower and seed in a short time following rainfall. In other species transpiration rate is reduced because of a waxy cuticle, few and sunken stomata, a surface of fine hairs and curled leaves. A number of species store water in succulent stems and leaves whereas others have either deep root systems below the water table or shallow roots which can absorb surface moisture.

12.6   Worked Examples

12.1

Read through the following account of kidney function and then write on the dotted lines the most appropriate word or words to complete the account.

Blood entering the kidney from the . . . . . . *renal artery* . . . . passes into an afferent arteriole which divides to form the . . . *glomerulus* . . . . inside the cup of a . . . *Bowman's capsule* . . . Much of the blood is forced into the tubule by the process of ultrafiltration. Only blood cells and large molecules, such as . *proteins,* . . . . and some fluid remain in the blood vessel. The filtrate is a watery fluid rich in food substances such as glucose and . *amino acids* . . . . Normally all the glucose is reabsorbed by the blood vessels surrounding the proximal tubule though a low level of the hormone . . . . . *insulin* . . . . may cause some to be excreted in the urine. Most of the . . *sodium ions* . . . . are also reabsorbed, causing a passive movement of most of the water out of the tubule due to the higher osmotic pressure now exerted by the blood. Further reduction in the water content of the filtrate takes place when . . *sodium ions* . . . . . . diffuse into the loop of Henlé and are later pumped out so producing a . *hypotonic* . . . . . . filtrate. Any increase in the . . *osmotic pressure* . . of the blood stimulates receptor cells in the . . . *hypothalamus* . . . . . of the brain which results in the secretion of . . *anti-diuretic hormone* by the pituitary gland. This hormone . . *increases* . . the permeability of the wall of the tubule so that. . *more* . . . . . . water leaves the tubule.

[14]
[L]

12.2

Figure 12.3 represents a mammalian kidney nephron.

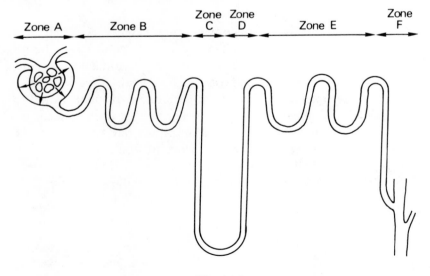

Fig. 12.3

150

(a) Name the process, indicated by the arrows, taking place in Zone A.

ultrafiltration

.... [1]

(b) Complete the table below to indicate (i) the approximate proportion of each of the listed substances that passes out of the tubule in Zone B, and (ii) the process by which each of the substances passes out of the tubule in Zone B.

Substances	(i) Approximate proportion of substance that passes out of the tubule in Zone B	(ii) Process by which the substance passes out of the tubule in Zone B
Water	85%	*osmosis*
Glucose	100%	*active transport*
Sodium ions	85%	*diffusion and active transport*

[6]

(c) In which zones of the kidney tubule does urea pass out of the tubule?

B and F

.... [2]

(d) The substances which leave the tubule pass into the extracellular fluid. Where do these substances go after that?

capillary bed of vasa recta

.... [1]

(e) (i) What passes out of the tubule in Zone D?

sodium ions

....

(ii) How does it leave?

active transport

....

(iii) Where does it go?

tissue fluid in medulla

....

(iv) Name the hormone that controls the amount leaving.

aldosterone

....

[4]

(f) Is the fluid which passes from Zone D to Zone E in the tubule hypertonic or hypotonic to the blood?

hypotonic

.... [1]

(g) What is the main substance which may leave the tubule in Zones E and F?

water

.... [1]

(h) Briefly explain the homeostatic mechanism controlling the functioning of parts E and F when the osmotic pressure of the blood rises.

Increased blood osmotic pressure detected by hypothalamus.

This stimulates the pituitary gland to secrete ADH. This

hormone increases the permeability of the tubule walls of

E and F and water is lost. This passes into the medulla

and into blood capillaries by osmosis. The volume of the

urine decreases and a hypertonic urine is produced.

[5]

[L]

12.3

(a) Complete the following table:

	Main nitrogenous waste substance	Main site of nitrogenous excretion
Freshwater protozoa	ammonia	plasma membrane
Adult insects	uric acid	Malpighian tubule
Freshwater fish	ammonia	kidney
Mammals	urea	kidney

(b) Many reptiles live in hot arid regions. How does their nitrogenous waste-product contribute to their survival in this habitat?

Reptiles excrete uric acid which is almost insoluble. This is removed from the cloaca as dry pellets. Little water is lost in this way.

[5]

[AEB]

12.4

(a) Table 12.1 shows the composition of the plasma, glomerular filtrate and urine.

Table 12.1

Substance	Concentration (g 100 cm^{-3})		
	Plasma	Glomerular filtrate	Urine
Water	90–93	97–99	96–97
Proteins	7–9	0	0
Creatinine	0.001	0.001	0.15
Glucose	0.1	0.1	0
Urea	0.03	0.03	2.0
Uric acid	0.002	0.002	0.05
Ammonia	0.0001	0.0001	0.05
Calcium	0.01	0.01	0.015
Magnesium	0.002	0.002	0.01
Potassium	0.02	0.02	0.15
Sodium	0.3	0.31	0.35
Chloride	0.35	0.37	0.6
Phosphate	0.003	0.003	0.12
Sulphate	0.003	0.003	0.18

(i) Identify the MAIN differences between the composition of the plasma and glomerular filtrate, and the glomerular filtrate and urine.

(ii) Describe the processes which bring about these differences. [8]

(b) Describe the hormonal control of the water content in the body of a mammal. [3]

[NISEC]

(a) (i) Plasma contains proteins whereas glomerular filtrate is protein-free. Urine contains greater concentrations of creatinine, urea, uric acid, ammonia, chloride, phosphate and sulphate, and a lower concentration of glucose, than glomerular filtrate.

(ii) As plasma passes through the glomerulus all molecules with a relative molecular mass less than 68 000 are forced out of the capillaries by the hydrostatic pressure of the blood. This process is called ultrafiltration and is aided by the fenestrations of the capillaries and the podocytes of Bowman's capsule. Proteins and blood cells cannot pass through these pores.

All the useful substances in the glomerular filtrate are selectively reabsorbed from the tubules and returned to the plasma. Only waste nitrogenous substances and mineral salts in excess of metabolic needs are not reabsorbed. These remain in the tubule and appear in the urine.

(b) The hormone aldosterone influences the passage of sodium ions from the distal convoluted tubule filtrate to the capillaries. An osmotically equivalent amount of water follows by osmosis, thus preventing excess water loss from the body.

Osmoreceptors in the hypothalamus respond to a rise in blood osmotic pressure by causing the release of anti-diuretic hormone (ADH) from the posterior pituitary gland. ADH increases the permeability of the walls of the distal convoluted tubule and collecting duct to water. Water is withdrawn from the filtrate into the vasa recta capillaries. When the blood osmotic pressure reaches a normal level the release of ADH is prevented by a negative-feedback process.

12.5

Figure 12.4 shows a transverse section through the leaf of a species of grass of the genus Festuca *as it would appear in dry air.*

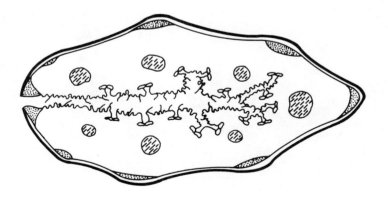

Fig. 12.4

(a) Describe three features visible in the diagram which identify the grass as a drought-resistant species.

(i) ....thick cuticle..

(ii) ....inward curled leaf....................................

(iii) ....sunken stomata......................................

(b) What drought-resistant feature would you expect the root system of this species of Festuca *to possess? Give reasons for your answer.*

....Deep, much-branched rooting system. The roots can penetrate....

....below the water table and a large surface area of root hairs....

....ensures maximum water extraction.........................

(c) State two features of xerophytes not shown in the diagram.

(i) ....reduced leaf size.....................................

(ii) ....water storage tissue.................................

[7]

[AEB]

154

13 Reproduction

Asexual reproduction   Reproduction by a single organism without the production of gametes and resulting in the production of genetically identical offspring.

Sexual reproduction   Production of offspring by the fusion of the genetic material of haploid nuclei.

Hermaphrodite   An organism capable of producing both male and female gametes.

Life cycle   The stages of development through which members of a species pass from the zygote of one generation to the zygote of the next.

Pollination   The transfer of the male gamete inside the pollen grain from anther to stigma.

In general terms reproduction is the passing of genetic material from parent to offspring. This ensures that the characteristics of the parent(s) and species are perpetuated.

13.1   Asexual Reproduction

This is more common in plants than animals and is achieved in a variety of ways.

(a) Fission

Binary fission (e.g. *Amoeba*) occurs when the parent cell grows to a particular size and then divides *mitotically* into two daughter cells genetically identical to the parent. Replication of DNA and nuclear cleavage precede cytoplasmic division.

   Multiple fission (e.g. *Plasmodium*) occurs when the parent cell nucleus divides *mitotically* repeatedly. Cytoplasmic division follows and numerous daughter cells are produced which offset the large fatalities of these cells as they transfer between vector and host.

(b) Sporulation (e.g. Bacteria, Fungi)

This is characteristic of saprophytic and parasitic micro-organisms. Vast numbers of **spores**, each containing a small amount of cytoplasm and a nucleus, are used to enhance the chances of survival and finding locations of new food sources.

(c) Budding (e.g. *Hydra*, *Bryophyllum*)

An outgrowth from the parent plant develops and eventually drops off to live an independent existence. It is a genetically identical copy of the parent.

(d) Fragmentation (e.g. *Spirogyra*, *Planaria*)

The parent organism fragments into two or more parts. Each of these is capable of developing into a new individual capable of living an independent existence.

(e) Vegetative Propagation

A highly differentiated region of the parent plant vegetative body separates from the parent plant and grows and develops into a completely independent plant. A number of structures which develop in this way are also able to store food. In this way these structures are organs of **perennation** as well, being able to survive adverse conditions for a considerable period of time. Such specialized structures include bulbs, corms, rhizomes, tubers, stolons, runners and swollen tap roots.

(f) Artificial Propagation

This process is used extensively by man when trying to produce new varieties of plants for food or ornamental use. Once the new variety has been produced, the aim is to avoid any variation in the offspring. Plant parts called **cuttings** are taken from the parent plant, given suitable growing conditions and allowed to develop new roots and leaves, thus becoming independent organisms.

Grafting occurs when a stem cutting from the donor plant (**scion**) is attached to the main stem of a rooted recipient woody plant (**stock**). The end-product is a plant with the root system of the stock and the shoot system of the scion. This method is exploited in the commercial growing of roses and apple trees.

13.2   Sexual Reproduction

In most species (except Bryophytes), **gametes** of two different sexes carrying genetic material derived by meiosis fuse during **fertilization** to form a diploid **zygote**. Meiosis and the random fusion of these gametes provides the genetic basis of variation within a species.

Parent organisms may be single-sexed (**dioecious**) or hermaphrodite (**monoecious**). Hermaphrodites may display self-fertilization or possess mechanisms which ensure outbreeding.

(a) Sexual Reproduction In Plants

The phenomenon of **alternation of generations** occurs in all terrestrial green plants. In this process a **haploid gametophyte** generation alternates with a **diploid sporophyte** generation. For the purposes of the following descriptions, knowledge of flower structure is assumed.

(i)   *Development of Pollen Grains*

Spore mother cells within the **anther pollen sacs** divide *meiotically* to form four haploid **pollen grains**. The nucleus of each grain then divides mitotically to form a **generative nucleus** and a **pollen tube nucleus**. At this point the pollen grain is equivalent to the male gametophyte.

(ii) *Development of the Embryo Sac and Female Gamete*

One diploid **embryo sac mother cell** develops within the **nucellus** of the **ovule**. It then divides *meiotically* and one of the cells produced becomes the haploid **embryo sac**. The embyro sac nucleus now divides *mitotically* to form eight nuclei, one of which is the **female gamete nucleus**. **Two polar nuclei** fuse to become a single diploid nucleus. The other five nuclei serve no further function.

(iii) *Pollination*

Self-pollination occurs when the **stigma** receives **pollen** from **stamens** of the same plant. It occurs in hermaphrodite flowers only, as a result of the simultaneous ripening of stamens and carpels. Self-pollination is independent of animals and wind. Because it is a method of inbreeding, it may result in the propagation of less vigorous offspring.

Cross-pollination is the transfer of pollen from the **anther** of one plant to the **stigma** of a flower of a different plant. It is a form of outbreeding and increases the chances of variation. However, it must rely on external agencies such as wind, water and animals (especially insects). Insect-pollinated (**entomophilous**) flowers in particular demonstrate a tendency for reduction and fusion of floral parts, protection of the ovary and bilateral symmetry instead of radial symmetry. Such adaptations for insect and wind pollination should already be fully understood.

(iv) *Fertilization*

Sucrose secreted by the stigma stimulates a compatible pollen grain to germinate and produce a **pollen tube**. This grows down through the style and locates an ovule in the ovary by chemotropism. Growth of the tube is stimulated by auxins from the gynaecium and is controlled by the pollen tube nucleus.

Within the pollen tube the generative nucleus undergoes mitosis to produce two non-motile **male nuclei** (gametes). As the pollen tube enters an ovule via its **micropyle**, the tip of the tube bursts. The two male nuclei enter the ovule, one fusing with the female gamete to form a **diploid zygote** and the other fusing with the diploid nucleus to form the **triploid primary endosperm nucleus**.

(v) *Seed Development*

A fertilized ovule is called a **seed**, and an ovary containing seeds is called a **fruit**. The zygote undergoes cell division to form a multicellular **embryo** consisting of a **plumule**, **radicle** and one or two **cotyledons** in mono- and dicotyledons respectively, which may act as food storage tissue. The primary endosperm nucleus divides to form the **endosperm** which in some seeds acts as a food store. As these developments take place the nucellus breaks down, supplying food for growth. The **integuments** become the thin, tough, protective testa. Finally, water is withdrawn from the seed, reducing metabolic activity and ensuring dormancy.

During seed development the ovary becomes a mature fruit, and protects the seeds and may aid their dispersal. Sometimes the receptacle is involved in fruit formation. Fruits formed in these cases are termed **false fruits** (e.g. strawberry). The remaining floral parts wither and die and finally fall away.

(vi) *Fruit and Seed Dispersal*

Fruits and seeds may be dispersed by wind, water or animals and display a variety of adaptations for this purpose. In addition, self-dispersal mechanisms exist involving the mechanical explosion of the seeds from the fruit. The further away the seeds are dispersed the less competition there will be with the parent plant but the less likely it is that new locations suitable for germination will be found.

(b) Sexual Reproduction In Man

It is assumed that the student will know the structure of the male and female reproductive systems.

(i) *Gametogenesis*

In males and females this process involves multiplication, growth and maturation phases.

(ii) *Spermatogenesis*

Sperms are produced by meiosis in the **seminiferous tubules** of the **testis**. The process takes about 70 days. Repeated division by the cells of the **germinal epithelium** produce **spermatogonia**. These increase in size and differentiate into **primary spermatocytes** which undergo two successive meiotic divisions to form **secondary spermatocytes** and **spermatids** respectively. Eventually the spermatids differentiate into mature **spermatozoa**. Fluid secreted by **Sertoli cells** bathes the sperms while they are in the tubules. It also provides protection and nourishment for them. Spermatozoa of all species share a common structure (see the answer to Worked Example 13.3).

(iii) *Endocrine Function of the Human Testis*

Between the seminiferous tubules are **interstitial cells** which secrete **testosterone**. This hormone is necessary for the successful production of sperm and also controls the development of male secondary sexual characteristics during puberty.

(iv) *Oogenesis*

Germ cells in the foetus produce many **oogonia** which divide mitotically to form **primary oocytes**. These are enclosed in groups of **follicle cells**. The primary oocyte divides meiotically into the haploid **secondary oocyte** and first **polar body**. Second meiotic division in the oocyte proceeds as far as metaphase but is not completed until the oocyte is fertilized by a sperm.

The **oestrous** or **menstrual cycle** refers to the cyclical changes which take place in the female reproductive system during the production of **ova**. They are controlled by **oestrogen**, **progesterone**, **luteinizing hormone** and **follicle-stimulating hormone** (see the answer to Worked Example 13.5). Following menstruation the **endometrium** thickens and becomes more vascular under the influence of oestrogen and follicle-stimulating hormone. During the first fourteen days of the oestrous cycle the **Graafian follicles** mature. **Ovulation** occurs on about day 14 and the secondary oocyte is expelled and swept into the **oviduct**. The ruptured follicle becomes the **corpus luteum**. This secretes progesterone which stimulates further thickening and vascularization of the endometrium, which also becomes fully secretory. If fertilization does not occur, the levels of oestrogen and progesterone decrease, causing endometrial breakdown and its expulsion in the menstrual flow (see the answer to Worked Example 13.5).

(v) *Fertilization*

Copulation occurs when the male inserts his erect **penis** into the **vagina** of the female and **ejaculates** sperms into the female reproductive tract. Before this occurs **seminal fluids** mix with the sperm to form **semen**. The fluids contain fructose for nourishment and prostaglandins which cause the female system to contract and assist the passage of the sperms towards the ovum. During their journey along the female reproductive tract the sperms undergo a process called **capacitation** which changes the properties of the **acrosome membrane** (see the diagram in the answer to Worked Example 13.3). When a sperm touches the outer membrane of a secondary oocyte the acrosome breaks to release hydrolytic enzymes which assist the sperm in penetration of the egg. Once the entire sperm has entered the oocyte the oocyte membrane changes into a **fertilization membrane** in order to prevent further sperm entry. Next the oocyte completes its second meiotic division, forming an **ovum** and **second polar body** (which degenerates). The ovum fuses with the sperm nucleus to form a diploid zygote.

The zygote divides mitotically and rapidly develops into a **blastocyst**. Its outer **trophoblastic cells** secrete **human chorionic gonadotrophin** which prevents autolysis of the corpus luteum and thus maintains progesterone secretion. The trophoblast ultimately gives rise to the foetal **chorion** and **amnion** membranes, and, as part of the **allanto-chorion**, contributes to placenta formation. After ten weeks the placenta secretes most of the oestrogen and progesterone necessary to maintain pregnancy and prepare the **mammary glands** for **lactation**.

At the **placenta** a large surface area of foetal and maternal tissues is formed for the exchange of oxygen, carbon dioxide, waste, nutrients and other materials. During the final month of development the foetus acquires antibodies from its mother. These provide the foetus with some passive immunity until its own immune system begins to function. The placental barrier also protects the foetal circulation from the higher blood-pressure of the maternal circulation, antibody attack and blood-group mismatching between foetus and mother.

(vi) *Birth*

Gestation normally takes 40 weeks. Just prior to birth the progesterone level in the mother falls and oxytocin is released. This causes the onset of uterine contractions, dilation of the **cervix** and rupture of the amniotic sac with the release of amniotic fluid. Uterine contractions push the baby headfirst downwards as the cervix fully dilates. The baby then passes out of the mother and its **umbilical cord** is ligatured. Within several minutes of the birth the uterus contracts dramatically and separates from the placenta. The placenta is then expelled as the 'afterbirth'.

(vii) *Lactation*

The initial secretion passed to the baby is **colostrum**, a mixture rich in protein and globulins. It serves as a means of passing antibodies to the baby. Prolactin stimulates milk production by the glandular cells of the mammary glands which is then passed to the nipples. Human milk contains fat, lactose and proteins and it is ejected by the milk ejection reflex.

13.3 Worked Examples

13.1

(a) *Name the parts indicated by the label lines on the diagram of the flower (Fig. 13.1).*

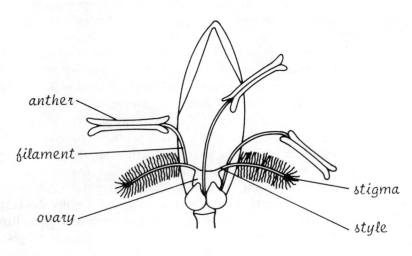

anther

filament

ovary

stigma

style

Fig. 13.1

[5]

(b) *Suggest the method by which this flower is pollinated.*
.... wind .... [1]

(c) *Give four features shown by this flower which support your answer to (b).*
 (i) ...... exposed anthers ....
 (ii) ...... large anthers ....
 (iii) ...... feathery stigma ....
 (iv) ...... stigma outside flower ....
[4]
[L]

13.2

(a) *In the space below, draw a large diagram of an unfertilized ovule of a flowering plant.*
     Label on your diagram the following parts:

 (i) micropyle   (iv) embryo sac (female gametophyte)
 (ii) nucellus   (v) polar nuclei
 (iii) integuments   (vi) female nucleus

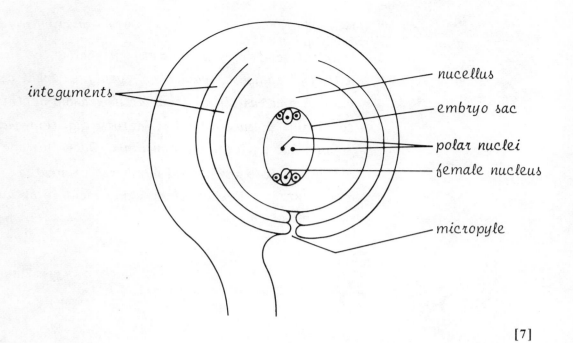

integuments

nucellus

embryo sac

polar nuclei

female nucleus

micropyle

[7]

(b) In the table, state three structural ways in which a wind-pollinated flower might differ from an insect-pollinated flower.

	Wind-pollinated flower	Insect-pollinated flower
(i)	small petals	large, coloured petals
(ii)	nectaries absent	nectaries present
(iii)	small, smooth pollen grains	large, sticky pollen grains

[3]

(c) In a species of flowering plant, the chromosome number of each cell in the radicle is 16. State the chromosome number of:

(i) the pollen tube nucleus . . . . 8 . . . .

(ii) an antipodal cell . . . . 8 . . . .

(iii) a cell of the endosperm . . 24 . . . .

(iv) a pollen mother cell . . . . 16 . . . .

(v) an integument cell . . . . . . 16 . . . .

[5]

(d) How do you account for the fact that machine-threshed legume seeds usually show a higher percentage of germination than those harvested by hand?

. . machine threshing breaks or damages testa and enables

. . water to enter for germination.

. . . .

. . . .

[2]

(e) How would you investigate the effect of sowing seeds at different depths in soil on the rate of germination?

Plant 10 seeds of same variety just below the surface of a tray of compost, spaced out evenly. Repeat using the same number and variety of seeds at increasing depths of 5 cm down to 25 cm. Ensure all other factors, e.g. temperature and water supply, are uniform and constant. Once 'surface' seeds germinate dig up the rest and record numbers germinated. Calculate as a percentage. Repeat whole experiment to obtain mean values of germination success.

[6]

[1]

13.3

(a) *(i)* Make a fully labelled diagram of a spermatozoon from a NAMED species.
(ii) What type of division produces spermatozoa? [4]

(b) Describe, with the aid of suitable diagrams, the process of fertilization as seen in vitro in a NAMED animal. [5]

(c) Describe the process of fertilization in an angiosperm from the moment of pollination. [3]

(d) Describe the development of the ovule into the seed in a NAMED angiosperm. [3]

[NISEC]

(a) (i)

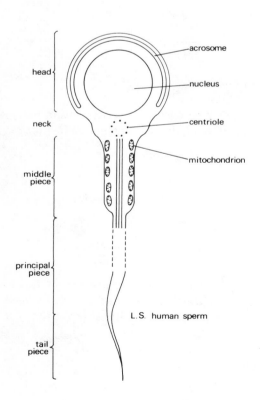

L.S. human sperm

(ii) Mitosis, between germinal epithelium cells and primary spermatocyte, followed by meiosis leading to spermatozoa production (a).

(b) When a sperm reaches the oocyte the membranes of the acrosome rupture, releasing hyaluronidase and protease enzymes. These digest the zona pellucida of the oocyte. This is called the acrosome reaction. In man, the entire sperm enters the oocyte (b). Cortical granules beneath the plasma membrane rupture and release a substance which causes the zona pellucida to thicken and separate from the plasma membrane. This is called the cortical reaction. It produces an impermeable fertilization membrane preventing polyspermy.

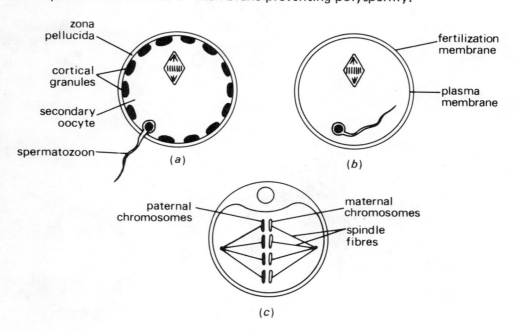

The second meiotic division of the oocyte now occurs, to produce the ovum and the second polar body. The latter and the sperm tail disappear. Sperm and ovum nuclei form pronuclei, their membranes break down and paternal and maternal chromosomes attach to spindle fibres (c). The diploid number of chromosomes is restored and the fertilized ovum is called a zygote.

(c) A pollen grain germinates on the stigma in the presence of a sucrose solution. The pollen tube grows down the style to the ovary. Its growth, and the secretion of digestive enzymes to facilitate its movement, are controlled by the tube nucleus of the pollen grain. Auxins produced by the gynaecium direct the pollen tube to the ovary.

During growth the generative nucleus of the pollen grain divides by mitosis to produce two male gametes. The pollen tube enters the ovule through the micropyle, the tube nucleus degenerates and the tip of the tube bursts, releasing the male gametes near the embryo sac. One nucleus fuses with the female gamete, forming a diploid zygote, and the other fuses with the two polar nuclei, forming a triploid nucleus, the primary endosperm nucleus.

(d) In the broad bean (*Vicia faba*) the zygote grows by mitotic divisions to become a multicellular embryo consisting of the plumule, radicle and two cotyledons. The cotyledons become swollen with food, which acts as storage tissue. The endosperm does not develop further, as its function in the broad bean is taken over by the cotyledons. Growth of all the tissues continues within the embryo sac, using nutrients from the nucellus and supplied by the funicle. The micropyle persists as a small pore in the integuments which gradually become tough and form the protective testa. The final stage of seed maturation involves a reduction in the water content of the seed.

13.4

The diagram (Fig. 13.2) is of a mammalian ovary.

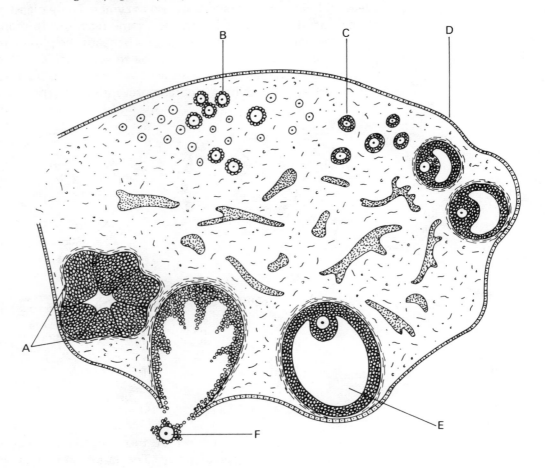

Fig. 13.2

(a) Which of the following structures does the letter A indicate? Underline the correct answer.
   (i) interstitial cells
   (ii) <u>corpus luteum</u>
   (iii) primary follicle
   (iv) secondary oocyte
   (v) germinal epithelium

[1]

(b) Which of the following is the correct sequence in the development of the labelled structures? Underline the correct answer.
   (i) A F E D C B
   (ii) E F B C D A
   (iii) C B D A F E
   (iv) <u>D B C E F A</u>
   (v) D A B C E F

[1]

(c) For each of the following, state whether it is haploid or diploid.
   (i) secondary oocyte ...... haploid ....
   (ii) primary oocyte ...... diploid ....

(iii) germinal epithelium .... *diploid* ....

(iv) oogonium .... *diploid* ....

[4]

(d) (i) Name one hormone produced by the ovary.... *progesterone* ....

(ii) What is the main role of this hormone?

.... *maintains the endometrium* ....

[2]

(e) State three general features of animal hormones.

(i) .... *secreted by endocrine glands* ....

(ii) .... *pass into blood-stream* ....

(iii) .... *exert specific effect away from gland* ....

[3]

(f) What part does the placenta play in the

(i) nutrition of the embryo .... *It forms a large surface area for the diffusion of nutrients from the maternal blood to the foetal blood. Glucose, amino acids and salts are transferred.* ....

[3]

(ii) protection of the embryo .... *The placenta isolates the foetus from the higher blood pressure of the mother and from direct connection of the two blood systems. Excretory materials can easily pass from foetus to mother.* ....

[3]

[L]

13.5

The diagram (Fig. 13.3) shows the blood levels of the hormones involved in the control of the human menstrual cycle: luteinizing hormone (LH), follicle-stimulating hormone (FSH), oestrogen and progesterone.

(a) For each hormone state

(i) where it is produced, and

(ii) the organ (or organs) on which it acts. [4]

(b) Confining yourself to a description of the effects of each of these hormones on the secretion of the others, show how negative feedback operates in this system. [2]

(c) Following menstruation, the uterine lining undergoes repair and then proliferation of secretory tissue. Name the hormones directly responsible for initiating

(i) repair, and

(ii) proliferation of secretory tissue. [2]

(d) Describe the development of the Graafian follicle from the oogonium to ovulation.

[4]

(e) Describe the process of implantation and formation of the placenta. [4]

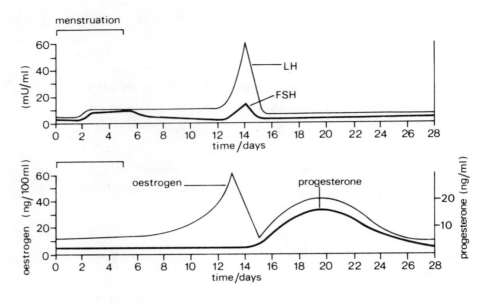

menstruation

time /days

time /days

Fig. 13.3

(f) *(i) Draw a simple sketch graph showing the changes in blood levels of oestrogen and progesterone in the event of fertilization and successful implantation occurring. You should continue your graph up to the point of parturition (birth).*

(ii) On the same sketch graph show the changes in level of one other named hormone involved in gestation and/or parturition. What is the function of the hormone you have named? [4]

[NISEC]

(a)

	(i)	(ii)
LH	anterior pituitary gland	ovary
FSH	anterior pituitary gland	ovary
oestrogen	ovarian follicle	uterus
progesterone	corpus luteum	uterus

(b) FSH stimulates growth of one of the ovarian follicles. When mature, the thecae of the follicle are stimulated by LH to produce oestrogen. These events occur between days 2 and 6. Increasing levels of oestrogen feedback negatively and cause a decrease in FSH levels without affecting LH levels. When oestrogen level reaches a maximum just before ovulation it has a positive feedback effect on the pituitary gland, causing a release of both FSH and LH.

As the LH level falls, the ruptured follicle becomes the corpus luteum and begins to secrete progesterone. As the level of progesterone rises the corpus luteum begins to secrete oestrogen. The high levels of oestrogen and progesterone inhibit further FSH release. At day 26 the corpus luteum begins to regress, oestrogen and progesterone levels fall and FSH and LH levels again begin to rise.

(c) (i) Repair: oestrogen.
(ii) Proliferation of secretory tissue: oestrogen and progesterone.

(d) A primordial germ cell undergoes repetitive mitotic divisions to produce an oogonium. An oogonium will undergo a further division to produce two primary oocytes. Each primary oocyte is enclosed by a layer of cells called the membrana granulosa and forms a primordial follicle. FSH causes the follicle to grow. The membrana produces an outer theca externa and an inner theca interna. The cells of the granulosa secrete a fluid which collects in the antrum. LH causes the thecae to produce oestrogen. By this stage the follicle is mature and known as a Graafian follicle. The primary oocyte then undergoes the first meiotic division to form the haploid secondary oocyte and the first polar body. At ovulation the follicle bursts and the secondary follicle is released.

(e) As the zygote passes down the fallopian tube, cleavage occurs and gives rise to the blastocyst consisting of an outer layer of cells, the trophoblast and the inner cell mass. The cells of the trophoblast make contact with the cells of the endometrium and become embedded in the latter. This occurs about six to nine days after ovulation. Immediately following implantation, the trophoblast differentiates into two layers. The outer layer forms trophoblastic villi which grow into the endometrium. The spaces between the villi form cavities called lacunae which makes the endometrium appear spongy. Hydrolytic enzymes released from the villi cause the blood vessels of the endometrium to break down and blood fills the lacunae. Meanwhile, the trophoblast continues to grow and forms a membrane called the chorion. An outgrowth from the embryonic hindgut comes into contact with the chorion and becomes richly vascularized by the formation of large chorionic villi to form the placenta. This structure then takes over the role of exchanging materials between the foetus and the mother.

(f)

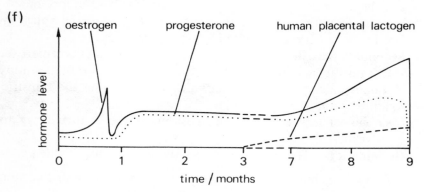

Human placental lactogen stimulates mammary development in preparation for lactation.

14 Growth and Development

Growth   An irreversible increase in the dry mass of protoplasm.

Cell differentiation   The developmental process by which a relatively unspecialized cell (or tissue) undergoes a progressive change to become a more specialized cell (or tissue).

Adventitious structures   Those structures growing in uncharacteristic positions, e.g. the adventitious roots that grow from the base of a stem.

Primary growth   Growth in plants originating in the apical meristem of shoots and roots which results in an increase in length.

Secondary growth   Growth in plants originating in lateral meristems (e.g. cambium) which results in an increase in girth and the production of woody tissue.

Metamorphosis   The changes which occur during the transition from the larval to adult form.

Cleavage   The successive cell divisions of the fertilized egg to form the multi-cellular blastula.

Germination   The resumption of growth by a spore, pollen grain or seed.

Growth in multicellular organisms occurs in three distinct phases: **cell division**, **cell expansion** and **cell differentiation**. When single-celled organisms divide this results in the growth of their population. Growth involves many biochemical activities controlled by enzymes. Changes in the overall form and structure of cells, organs and organisms are collectively known as **morphogenesis**. External factors such as light, heat, water and availability of food, and internal factors such as hormones interact with the DNA of the organism and therefore influence the overall growth process.

Differentiation is brought about by the differential expression of the genes. Cytoplasm also plays a role by repressing and activating genes in different cells. Hormones are also able to switch genes on or off, thus determining the pattern of development.

14.1   Measurement of Growth

A **sigmoid curve** is obtained when the growth of a measurable parameter is plotted against time. The curve consists of:

1. a **lag phase** where little growth is occurring,
2. a **log phase** where growth proceeds exponentially and growth rate is maximal,
3. a **decelerating phase** where growth begins to decrease being limited by internal and/or external factors,
4. the **plateau phase** where overall growth has ceased.

Some monocotyledonous leaves, invertebrates and fish exhibit *positive growth* until death, while man demonstrates *negative growth*, this being a corollary of senescence and ageing.

Plotting data from physical parameters (e.g. leaf length, body weight) against time produces an **absolute growth curve** which shows the overall growth pattern and the extent of growth. Plotting the change in parameter per unit time gives an **absolute growth rate curve** which shows how the rate of growth changes while the parameter is being measured. Dividing the absolute rate of growth by the amount of growth at the beginning of each time period provides a **relative rate of growth curve** which indicates the efficiency of growth.

14.2 Patterns of Growth

Different parts of a plant usually grow at different rates from each other and from the overall rate of the body. This produces a change in the size and form of the organism and is called **allometric growth**. It also occurs in many animals, including man. **Isometric growth** takes place when an organ grows at the same mean rate as the rest of the body of the organism. This occurs in fish and some insects, e.g. locusts.

Limited growth is demonstrated by annual plants and most animals. After maximum growth the organism matures and undergoes negative growth during senescence which is followed by death. Woody perennials, invertebrates, fishes and monocot leaves exhibit unlimited growth by continuing to grow each year. Arthropods grow periodically (discontinuous growth). They can only increase in size when their confining exoskeleton has been shed and before the new one has hardened.

14.3 Growth and Development in Plants

(a) Germination

Germination of the seed begins usually after a period of **dormancy**. All seeds require the *external* factors water, oxygen and a suitable temperature for the process to proceed. Water increases the metabolic activity within seeds and food reserves are digested hydrolytically in the food storage centres. The soluble products are translocated to the growing regions for utilization. Germination is characterized by a high metabolic rate and the synthesis of proteins which are used to make enzymes and structural components of the protoplasm. Glucose is used to make cellulose for the walls of newly produced cells. Gibberellin and cytokinin are both involved in the control of germination.

If the cotyledons are carried above the ground, germination is said to be **epigeal**. **Hypogeal** germination occurs if the cotyledons remain below ground. Primary growth is confined to apical meristems and ultimately results in the formation and elongation of roots, stems, lateral branches and the differentiation of specialized tissues.

(b) Primary Growth of the Root

Groups of **meristematic cells (initials)** give rise to new parenchymatous root cap cells which protect the apex as the root grows through the soil. These cells contain starch grains which act as **statoliths** and play a part in the root's response to gravity. Behind the root initials, cells differentiate into epidermis, parenchyma and the central vascular cylinder respectively. In the region of differentiation, root hairs develop from the epidermis. Lateral roots are formed when cells in the pericycle resume meristematic activity. The lateral roots are thus said to arise endogenously and form a new root apical meristem. **Adventitious roots** develop completely independently of the primary root and frequently form the rooting systems of monocotyledons, arising from stem nodes.

(c) Primary Growth of the Shoot

Immediately behind the shoot apex, before any differentiation occurs, groups of cells arise which will later develop into lateral buds, leaves, epidermis, ground tissues and vascular tissues. In the **zone of elongation** the cells take in water and their small vacuoles coalesce to form one large vacuole. Turgor pressure and the orientation of the cellulose microfibrils in the cell wall determine the final shape of the cell. Differentiation begins when cell expansion is nearly complete, with different cells laying down different amounts of extra cell wall materials and different shapes being assumed.

Protoxylem and **protophloem** are laid down first, to be followed by **metaxylem** and **metaphloem** in the **zone of differentiation. Vascular cambium** remains between the xylem and phloem and will later contribute to **secondary thickening**.

Development of lateral shoots is under the control of auxins from the shoot apex and the relative amounts of circulating auxin and cytokinin.

Phytochrome-controlled responses straighten the plumule on its exposure to light and change its etiolated condition to one of normal growth (**photomorphogenesis**). Now photosynthesis begins and the green plant assumes an independent autotrophic existence.

(d) Secondary Growth in Woody Dicot Stems

Secondary growth is the result of meristematic activity of both the **fascicular** and **interfascicular cambia**. Secondary phloem is laid down outside the cambium, and secondary xylem inside the cambium. In order to keep pace with the increase in girth, radial divisions of the cambial cells occur. **Primary medullary rays** run from the pith to the cortex, maintaining a living link between the two tissues by translocating water and salts from the xylem and soluble nutrients from the phloem radially across the stem. They also store food during dormancy.

As the girth of the stem increases the epidermis ruptures. In response to this, and prior to the rupture of the epidermis, a **cork cambium** or **phellogen** develops which produces cork cells to the outside and parenchyma inside. The cork cells become blocked with impermeable suberin and eventually die. **Lenticels** occur randomly scattered in the cork and permit gaseous exchange to take place between the external atmosphere and the living stem tissues.

As the woody dicotyledon grows older, the xylem cells in the centre of the trunk cease to function. They become blocked with tannins and this region is called **heartwood** as distinct from **sapwood**, which consists of the peripheral functional xylem cells.

When growth resumes each spring, large thin-walled cells are produced which conduct water to initiate expansion of newly formed cells. In the autumn fewer, narrower, thick-walled vessels are produced. The difference between the shape and size of the spring and autumn ring can be easily seen and constitutes the annual rings of the trees.

14.4 Metamorphosis

The changes that take place during metamorphosis adapt the organism for adult life generally in a new habitat. Details of the process are provided in the answer to Worked Example 14.4.

14.5 Development in Vertebrates

The development of the zygote until the time of hatching or birth is a continuous process. However, for ease of learning and understanding it can be divided into three stages:

(a) **Cleavage** is a series of *mitotic* divisions of the zygote into a hollow **blastula** enclosing a **blastocoel cavity**. Cleavage increases the nucleus : cytoplasm ratio and therefore increases nuclear control over the cytoplasm.

(b) **Gastrulation** involves the movement of cells into new positions and establishes the axes of the embryo and three **primary germ layers**, the ecto-, meso-, and endoderm. A blastopore appears and a segment of its dorsal lip, called the 'organiser', induces the cells overlying it to form the neural tube.

(c) **Organogeny** next occurs, producing definitive tissues by embryonic induction, organs and organic systems. Further cell division, growth and differentiation occurs. This is curtailed by hatching or the birth of the young individual.

The whole developmental process is regulated by hormones secreted by the thyroid, liver, adrenal cortex and gonads. These glands are themselves under the influence of other hormones released from the pituitary.

14.6 Worked Examples

14.1

(a) *'True growth is not simply an increase in size.' State below four different ways in which true growth may be defined.*

(i) *increase in dry mass* ....

(ii) *increase in cell number* ....

(iii) *irreversible increase in volume of cytoplasm* ....

(iv) *increase in differentiation* ....

[4]

(b) *State two external factors which influence growth in plants and describe one effect of each.*

(i) factor *light intensity* ....

effect *influences rate of photosynthesis* ....

(ii) factor *temperature* ....

effect *influences metabolic rate via enzyme action* ....

[4]

171

(c) Fill in the spaces in the following table which refers to hormones involved in growth processes.

Name of hormone	Site of hormone production	Effect
thyroxine	thyroid gland	controls basal metabolic rate
follicle-stimulating hormone	anterior pituitary gland	maturation of Graafian follicles
auxins	stem apex root apex	cell elongation
Gibberellins	all young plant tissues	stimulates cell growth

[8]
[L]

14.2

Seedlings from 100g of maize seeds were grown in the dark for 10 days. The seedlings were then analysed and compared with 100 g of ungerminated maize seeds. The following results were obtained.

	Dry mass of ungerminated seeds	Dry mass of seedlings after 10 days
Cellulose	2 g	5 g
Starch	63 g	9 g
Other organic material	13 g	27 g
Ash	2 g	4 g
Total dry mass	80 g	45 g

(a) Why is dry mass used for comparison?

It indicates the amount of organic material present which is a measure of change in mass of cytoplasm. [1]

(b) How would one ensure that the drying process had been completed?

Weigh, reheat at $110^{\circ}C$ for several hours. Cool. Constant mass. [1]

(c) Account for the decrease in the total dry mass of the seedlings.

Most of mass is starch which is converted to sugars and used up in respiration and other metabolic activities. [1]

(d) Why did the seedlings contain more cellulose than the ungerminated seeds?

Cellulose is synthesized during growth of new cell walls. [1]

(e) What is the most likely source of the carbon used to form this new cellulose?

Starch ⟶ glucose ⟶ cellulose [1]

[SEB]

172

14.3

Distinguish concisely between growth, differentiation and development. How would you measure growth in (a) a culture of bacteria; (b) a potted plant; (c) a small mammal?

The following values were obtained for the dry weights of germinating potato tubers and their sprouts:

Week	1	2	3	4	5	6	7
Total dry wt. (g)	54.0	48.8	44.4	34.8	36.4	46.4	71.2
Stalks, dry wt. (g)	0.0	0.0	0.0	1.6	2.8	10.0	23.2
Leaves, dry wt. (g)	0.0	0.0	0.0	0.8	2.8	12.4	26.8

Examine the changes of weight week by week and comment on their causes.

[OLE]

Growth is the result of metabolic activities increasing the dry mass of protoplasm. Differentiation is a change in the form or activity of a group of cells to enable them to perform a more restricted group of activities more efficiently. Development is the overall morphogenic change in an organism resulting from growth and differentiation.

(a) Set up a sterile flask containing 50 cm^3 of sterile nutrient broth. Inoculate with a source of a bacterium. Aerate the nutrient with sterile oxygen and maintain at 37 °C in a water-bath. At intervals of four hours, shake the flask carefully and aseptically remove a known volume of broth. Place this on a haemocytometer slide and count the bacteria present in a known area. Repeat this procedure at least four times to obtain a mean value. After twenty-eight hours plot a graph of numbers of bacteria against time.

(b) Take a potted plant such as a *Pelargonium* and connect it to an auxanometer. This lever device has a pointer attached to a scale. The relative movement of the pointer across the scale indicates the rate of growth of the plant. This method is crude but can give an indication of the rate of growth of the stem internodes.

(c) Growth in a mouse can be measured by measuring the length of its tail and plotting these lengths against time. The mouse is placed carefully into a special box which has an opening for its tail. The tail then rests on a millimetre scale and the readings can be taken directly. It is important to always use the same measuring box. By graphing total length against time the rate of growth of the tail can be determined. These data provide a good indication of the overall growth process.

Weeks 1, 2 and 3

More mass is lost between weeks 1 and 2 than between 2 and 3. These losses are due to starch conversion to glucose and its subsequent utilization as a respiratory substrate.

Weeks 3 and 4

The total dry mass decreases faster and more starch is used up in increased respiration which is needed to meet the metabolic demands of the developing stalks and leaves. The stalks grow faster than the leaves.

Weeks 4 and 5

The total dry mass now increases as the leaves have developed sufficiently to begin to photosynthesize. New starch is being synthesized and the stalks increase in mass.

173

Weeks 5 and 6

The total dry mass again increases due to photosynthetic activity. Both stalks and leaves continue to grow rapidly; the stems elongating and new leaves forming.

Weeks 6 and 7

The greatest increase in total dry mass per week has occurred. By now the plant will be well developed and starch may now be stored in tubers.

14.4

(a) Give a brief account of the growth and development, from egg to adult, of a named insect. [6]
(b) What is the significance of the larval phase in the life cycle of insects? [3]
(c) Describe the role of hormones in insect metamorphosis. [9]

[UCLES]

(a) Housefly (*Musca domestica*) eggs are laid in batches of 100 eggs on organic debris. After about twenty-four hours the eggs hatch and produce very small larvae which feed almost continually. Over the next four days the larvae grow and shed their cuticles. The formation of the new cuticle is called ecdysis and the shedding of the skin is called moulting. The stages in between moults are called instars. During these times cells multiply very rapidly. The cuticle of the last larval instar darkens and forms the pupa. During the non-feeding, apparently dormant pupal stage differentiation of tissues occurs so that on moulting of the pupa the imago emerges. Such a life cycle as shown by the housefly is called holometabolous (complete metamorphosis).

(b) Insect larvae have different feeding habits from adults and this enables them to exploit other food resources from the parents, therefore overcoming problems of competition. Many insects overwinter as the pupa stage. In this state of diapause their physiological activity is minimal. Consequently they are less susceptible to environmental extremes. This has survival potential. Finally, the presence of larval stages as the phases of growth allows the adult form to adapt to its major role of reproduction free from the structural adaptations associated with rapid growth.

(c) Insect metamorphosis is regulated by environmental factors which influence hormonal and nervous activity within the insect. Food availability and certain light and temperature conditions influence neurosecretory cells in the brain to release prothoracicotrophic hormone (PTTH) which passes down axons to be stored in the corpus cardiacum. The stimulus to moult may vary from species to species but in all cases it causes the release of PTTH into the blood. PTTH acts on two further glands, the corpus allatum and the prothoracic glands. The prothoracic glands secrete a moulting hormone called ecdysone. The corpus allatum produces another hormone called juvenile hormone. The simultaneous release of both hormones influences the formation of new cuticle to produce larval cuticle and larval body form. As metamorphosis proceeds the amount of ecdysone released falls off. At a critical level of ecdysone/juvenile hormone, a pupal ecdysis occurs. At the time of the next moult no more juvenile hormone is released and ecdysone causes the development of the adult form.

It has been demonstrated that ecdysone influences a particular stage of development by acting directly on genetic transcription. Juvenile hormone exerts its effects by differentially modifying this response, according to its relative concentration with respect to ecdysone.

15 Chromosome Behaviour

> **Chromosome**   A structure composed of DNA and protein found in the nucleus. Specific regions of the DNA are called genes and these are arranged in a linear order along the length of the chromosome.
> **Mitosis**   The process by which a cell nucleus divides to produce two daughter nuclei containing identical sets of chromosomes to the parent cell.
> **Meiosis**   The process by which a cell nucleus divides to produce four daughter nuclei each containing half the number of chromosomes of the original nucleus.
> **Mutation**   A sudden inherited change in the amount, or structure, of genetic material (DNA).
> **Polyploidy**   The possession of more than two complete sets of chromosomes per cell.

15.1   Chromosome Structure

Chromosomes are composed of **DNA**, which takes the form of a long **double helix** which in places forms aggregates with nucleoprotein histones. These structures are called **nucleosomes** and have the appearance of beads on a string. Details of the nature of attachment of the protein to DNA are still unknown. The DNA–protein complex is called **chromatin**. Small amounts of chromosomal RNA may also be present. For details of DNA structure see section 2.2.

When cell division takes place, cells pass through a regular sequence of stages known as the **cell cycle**. This includes two overlapping events, nuclear division followed by cell division.

15.2   Mitosis

Mitosis results in an increase in cell numbers and is the method by which growth, replacement and repair of cells occurs in all higher organisms. It provides an equal distribution of duplicate chromosomes between daughter cells and the production of **genetically identical** offspring. The overall process is continuous but can be conveniently broken down for ease of learning into a number of distinct phases. The length of time mitosis takes depends on external conditions such as temperature, food availability and oxygen.

(a) Interphase

The cell increases in size and synthesizes many new organelles. Just prior to the next phase, DNA and histones replicate. At this stage each chromosome exists as a pair of **chromatids** joined together by a **centromere**.

(b) Prophase

This is the longest phase. The chromatids shorten and thicken by **spiralization** and **condensation**. In animal cells and lower-plant cells the centriole pairs move towards opposite poles and a spindle forms. **Spindle fibres** are formed, made up of **microtubules** of tubulin. Nucleic acids from the nucleoli pass to the chromatids. Towards the end of prophase the spindle consists of long continuous fibres stretching from one pole to the other and shorter ones attached to the chromosome centromeres. The **nuclear membrane** finally breaks down, marking the end of prophase.

(c) Metaphase

Pairs of chromatids (chromosomes) become attached to spindle fibres at their centromeres and are manoeuvred into a position across the **spindle equator**.

(d) Anaphase

The centromeres *divide* and the chromatids (now chromosomes) separate. They are slowly pulled apart, centromere first, towards opposite poles. Therefore *identical* sets of chromosomes have been moved to each end of the spindle.

(e) Telophase

At the poles the chromosomes uncoil and elongate. The spindle fibres disintegrate and each mature centriole replicates. The nuclear membrane reforms and cytoplasmic cleavage begins. In animal cells **microfilaments** at the equator draw in the cytoplasm producing a furrow around the cell's circumference. The cell membranes meet and cell separation occurs. **A cell plate** is formed in plants by vesicles produced by Golgi bodies. **A primary cell wall** is laid down on either side of the cell plate, followed frequently by a **secondary cell wall**. Perforations in the wall house **plasmodesmata** which connect adjacent cells together (see section 3.3).

15.3 Meiosis

This occurs during **gametogenesis** and **spore production**. Alternatively called **reduction division**, it reduces the diploid number of chromosomes in a cell to a haploid number. This ensures that the chromosome number in offspring of sexually reproducing organisms is retained at a constant level from generation to generation. Essentially, the process consists of a duplication of genetic material (as in mitosis) in the parental cell followed by *two* successive meiotic nuclear divisions and cell divisions to form **four haploid cells**.

(a) Interphase

The cell increases in size and many organelles and the chromosomes replicate.

(b) Prophase I

Chromosomes condense. **Homologous** maternal and paternal **chromosomes** pair up (**synapsis**) and form structures called **bivalents**. The chromosomes coil around each other towards their ends while the centromeres appear to repel each other. **Chiasmata** may occur. These are the sites of exchange of genetic material during the process of **crossing over** and account for the major source of genetic variation. Bivalents assume a variety of shapes according to the number of chiasmata there are. The centrioles, if present, move to opposite poles, a **spindle apparatus** forms and the nucleoli and nuclear membrane break down. With bivalent formation the chromosome number appears to have been halved.

(c) Metaphase I

Each homologue of a bivalent is attached to the spindle by its centromere and aligned across the **spindle equator** with each centromere equidistant above and below it.

(d) Anaphase I

Spindle fibres pull the homologous chromosomes apart by their centromeres which do not divide. Thus *two haploid* sets of chromosomes are created.

(e) Telophase I

Nuclear membranes reform around the chromosomes, forming two new haploid nuclei. In some organisms this is followed by interphase II, while in others telophase I and interphase II are omitted and anaphase I passes straight on to prophase II.

(f) Prophase II

This stage is absent in cells without interphase II. The nucleoli and nuclear membranes disappear and new spindle fibres appear at right angles to the first spindle axis. The chromatids condense and centrioles, if present, migrate to opposite poles.

(g) Metaphase II

The chromosomes are *aligned* on the **equatorial plate** and attached to the spindle fibres by their centromeres.

(h) Anaphase II

The centromeres divide and the spindle fibres pull the separating chromatids (now called chromosomes) towards opposite poles.

(i) Telophase II

Spindle fibres disappear, centrioles replicate and nuclear membranes re-form. Chromosomes lengthen and uncoil and cell division proceeds as in mitosis. *Four haploid daughter cells* (gametes) are formed. They contain mixed paternal and maternal genetic material. This has occurred as a result of crossing over during meiosis and independent assortment of bivalents (see section 16.1).

15.4 Chromosome Mutations

Mutations occur when there is a change in number or structure of the chromosomes.

Changes in number arise when errors occur during meiosis. The loss or gain of chromosomes is called **aneuploidy**. It occurs when a pair or pairs of homologous chromosomes fail to separate (**non-disjunction**). Therefore the gametes contain one or more chromosomes too many or too few. Cells with an extra chromosome may be viable, producing a zygote with an odd number of chromosomes after fertilization. Down's syndrome in man ($2n = 47$) is one example of this mutation (see Fig. 15.7).

Polyploidy (euploidy) is the gain of whole sets of chromosomes. It is more common in plants than animals and is associated with **hybrid vigour**. **Autopolyploidy** occurs when the chromosome number of an otherwise normal individual becomes multiplied. This can occur when cytoplasm fails to separate during cytoplasmic cleavage.

Allopolyploids are formed by the hybridization of two closely related plants with different sets of chromosomes. Non-disjunction has to occur to produce somatic doubling of the chromosomes from which fertile gametes can be produced. The resulting organism will be fertile with polyploids like itself but not with its diploid parental species.

Structural chromosome changes produce a change in allele sequence and produce recombinants. During **inversion**, part of a chromosome breaks off, rotates 180° and rejoins the chromosome. The gene sequence is reversed. **Translocation** occurs when part of a chromosome breaks off and rejoins at the other end of the same chromosome or another non-homologous chromosome. **Deletion** is caused when a chromosome segment is lost altogether. Some chromosome regions **duplicate**, with the additional set of genes being incorporated within the parent chromosome or attached to another chromosome.

15.5 Gene Mutations

A gene mutation, as shown by a sudden and spontaneous change in phenotype, is the result of a change in the nucleotide sequence of the DNA molecule. Because this change occurs at a particular region of the chromosome it is sometimes called a **point mutation**. The change may involve the addition, loss or rearrangement of bases in the gene. Thus the mutations may take the form of the duplication, insertion, deletion, inversion or substitution of bases. In all cases the change in the nucleotide sequence will produce a modified polypeptide.

Most minor gene mutations are not seen in the phenotype since they are recessive. **Sickle cell anaemia**, however, an example of base substitution, has a profound effect on the phenotype. The substitution of *one* amino acid (valine) in place of another amino acid (glutamic acid) in one chain of 146 amino acids results in the production of sickle cell haemoglobin. Patients suffering from this mutation have acute anaemia and those with the homozygous recessive condition may die early from heart or kidney failure. In certain parts of the world where malaria is rife it is a selective advantage to carry the mutant in the heterozygous condition.

15.6 Worked Examples

15.1

(a) *Figure 15.1 shows the sequence of nitrogenous bases on part of a strand of DNA and the corresponding region of a strand of messenger RNA. Using the letters provided in the key, indicate on the diagram the complementary bases which would be found on the strand of messenger RNA.*

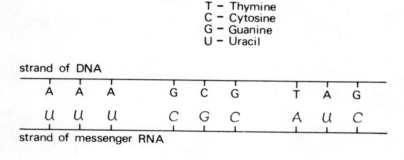

Key   A – Adenine
           T – Thymine
           C – Cytosine
           G – Guanine
           U – Uracil

strand of DNA

| A | A | A | | G | C | G | | T | A | G |

| U | U | U | | C | G | C | | A | U | C |

strand of messenger RNA

Fig. 15.1

[3]

(b) *State briefly why it is thought that the arrangement of bases of nucleic acids (DNA or RNA) is in the form of a 'triplet code'.*

Only a code of three bases could incorporate twenty amino acids into the structure of protein molecules. Also Crick's studies on frame-shifts in T_4 phages indicated that the code was non-overlapping in triplet bases.

[3]

(c) (i) *What is an alteration in the sequence of bases called?*

gene mutation

[1]

(ii) *Give one example of how this alteration might be caused.*

substitution of one base by another, e.g. in sickle cell anaemia

[1]

(d) *Name two groups of organisms in which the DNA is not enclosed within a distinct nuclear membrane.*

first group   bacteria

second group   blue-green algae

[2]
[L]

15.2

Using any of the terms from the following list, complete the diagram (Fig. 15.2) which illustrates the sequence of events in the addition of an amino acid to a peptide chain in the process of protein synthesis:

amino acid; anti-codon; centriole; chromatin; histamine; mitochondrion; mRNA; RNA; tRNA. **[2]**

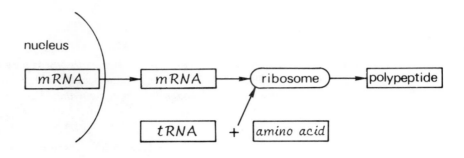

Fig. 15.2

[OLE]

15.3

Describe the mechanism of protein synthesis in the living cell. **[14]**

[UCLES]

DNA is composed of two polynucleotide chains associated so as to form a double helical molecule. The sequence of triplet bases in the polynucleotide chain carries the code for the synthesis of protein molecules. The sequence carrying the code for the production of a polypeptide is called a gene. As a preliminary to protein synthesis the hydrogen bonds between complementary strands of DNA break exposing a single strand of DNA which acts as a template for the formation of a complementary strand of messenger RNA (mRNA). This process is called transcription. Free nucleotides form mRNA according to the rules of complementary base pairing and in the presence of the enzyme RNA-polymerase. Once formed, the strand of mRNA leaves the nucleus through a nuclear pore and enters the cytoplasm where it attaches to a number of ribosomes. Here the sequence of mRNA bases is translated into a sequence of amino acids in a polypeptide chain.

Transfer RNA (tRNA) molecules pick up specific amino acid molecules from the cytoplasm according to their anti-codon triplet base sequences. While ribosomes hold the mRNA molecule, a specific tRNA–amino acid complex binds to the exposed triplet on the mRNA according to the rules of complementary base pairing. As further mRNA triplets are exposed, more tRNA–amino acid molecules are brought to the ribosome and peptide bonds form between adjacent amino acids. In this way new amino acids are added to the growing polypeptide chain to produce a primary structure determined in the nucleus by the sequence of triplet bases in the DNA.

Ribosomes will 'read' and 'translate' the mRNA code until they come to a codon signalling 'stop'. At this point translation is complete and the polypeptide leaves the ribosomes.

15.4

Figure 15.3 shows two cells from the same animal in the process of division.

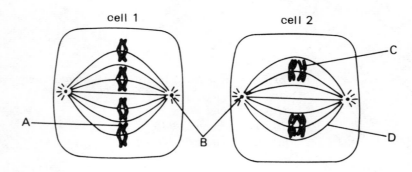

Fig. 15.3

(a) Name the structures labelled A to D. Write your answers on the dotted lines below.

A ..... chromatid .....

B ..... centriole .....

C ..... chromatid .....

D ..... spindle fibre .....

[4]

(b) (i) What type of division is shown in cell 1?

..... mitosis .....

(ii) What stage of division is shown in cell 1?

..... early anaphase .....

[2]

(c) (i) Describe the differences shown by the two cells.

1. homologous chromosomes are in pairs in 2 not in 1

2. chromatids are separating in 1, not in 2

[2]

(ii) What is the significance of these differences?

When homologous chromosomes separate in cell 2 the chromosome number is halved, not so in cell 1.

Identical chromatids separate in mitosis as a result of DNA replication, not so in cell 2, crossing-over possible.

[2]

(d) Which type of division occurs in a meristematic cell of a flowering plant?

..... mitosis ..... [1]

(e) *State the generation of a fern in which the type of division shown in cell 2 does not occur.*

gametophyte .... [1]

[L]

15.5

(a) *For a cell with two pairs of chromosomes, draw diagrams in the spaces indicated, to show*
  (i) *mitotic metaphase,*
 (ii) *meiotic metaphase I,*
(iii) *meiotic metaphase II.*

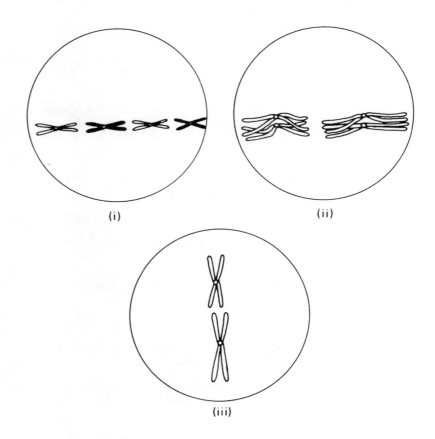

(i)   (ii)

(iii)

(b) *Complete the table below, using a 'plus' (+) to indicate presence and a 'minus' (−) to indicate absence during the period of nuclear division indicated.*

	Mitosis	*Meiosis I*	*Meiosis II*
Synapsis	−	+	−
Anaphase	+	+	+
Replication of DNA	+	+	−
Segregation of allelomorphic genes	−	+	−

[7]

[UCLES]

15.6

Figure 15.4 is a drawing of a bivalent at a particular stage of meiosis.

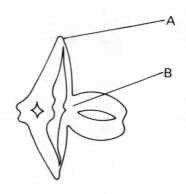

Fig. 15.4

(a) What is a bivalent?

..... *a pair of homologous chromosomes paired during meiosis*

(b) Name the structures at the positions marked A and B.

A ..... *centromere*

B ..... *chiasma*

(c) Make a labelled drawing below of how you would interpret the chromatid structure of the bivalent.

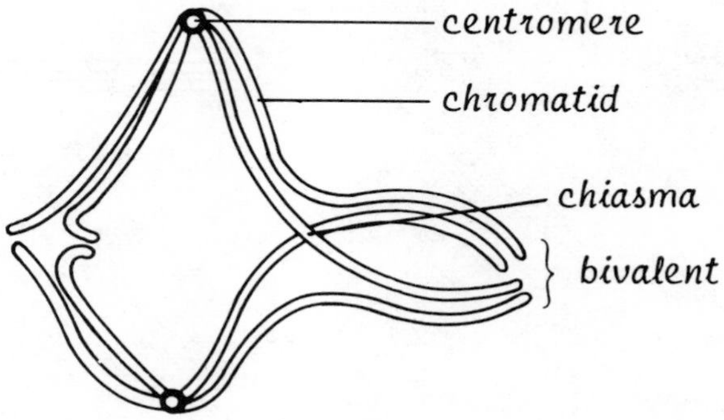

centromere

chromatid

chiasma

} *bivalent*

(d) At what stage of meiosis is the bivalent? Give your reasons.

Stage ..... *early anaphase I*

Reasons ..... *centromeres have moved apart and yet chiasma*

is still observable.

[6]

[UCLES]

15.7

(a) A and B, shown in Fig. 15.5 below, are dominant alleles of two genes situated on a pair of homologous chromosomes. The respective recessive alleles are shown by a and b.

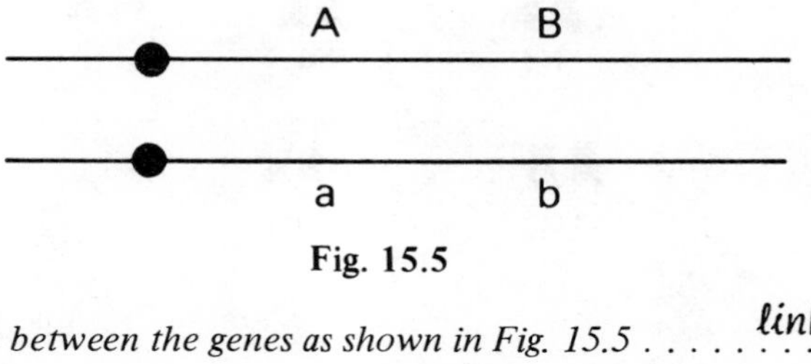

A   B

● ────────────────

● ────────────────

a   b

Fig. 15.5

Give the relationship between the genes as shown in Fig. 15.5 ..... *linked* ..... [1]

(b) Fig. 15.6 below represents the result of crossing over between the pair of homologous chromosomes depicted in Fig. 15.5.

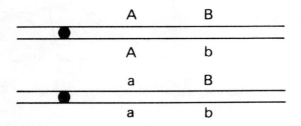

Fig. 15.6

Make a drawing or drawings in the space below to show how these two chromosomes will now behave on the spindle in early anaphase 1 of meiosis.

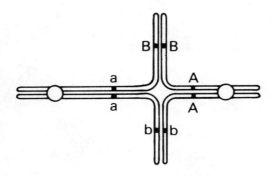

[OLE]

15.8

Figure 15.7 represents a human karyotype (a display of matched chromosomes).

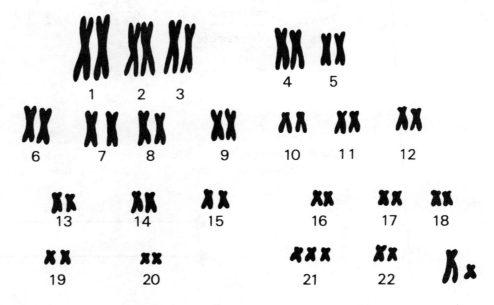

Fig. 15.7

(a) *Name the condition caused by the chromosome set 21 shown above.*

Down's syndrome

..... [1]

(b) *How does the condition shown by chromosome set 21 arise?*

Failure of a pair of homologous chromosomes to separate during anaphase II. One of the gametes will contain 2 of the type 21 chromosomes. At fertilization this combines with an extra type 21 chromosome.

[2]

(c) *Select one other feature of the karyotype and state the effect on the phenotype.*

Feature: presence of both X and Y chromosomes

Effect: confers male features on the phenotype

[2]

[SEB]

15.9

The plant genus Brassica *(Cruciferae) contains a number of species from which some of our commonest vegetable and fodder plants e.g. cabbage, swede and rape have evolved. Hybridization between the turnip (Brassica rapa, 2n = 20), and the black mustard (Brassica nigra, 2n = 16) has produced the brown mustard (Brassica juncea, 2n = 36).*

(a) *How are hybrid plants produced experimentally?* Remove unripe anthers from homozygous species and cover flower with muslin bag. Transfer pollen from another homozygous species to ripe stigma of first plant. Cover again.

[2]

(b) *What chromosomal change occurred during the formation of the brown mustard hybrid, and what term is used to describe such a hybrid?*

Non-disjunction of F_1 hybrids produces diploid gametes (2n = 18) which form F_2 hybrids (2n = 4x = 36). This is called allopolyploidy.

[2]

(c) *Why are plants of brown mustard able to produce fertile seed?*

Homologous pairing can occur during meiosis and diploid gametes are formed having 9 chromosomes from each parent.

[1]

(d) *A black mustard plant with 32 chromosomes has been produced but it has proved to be almost totally sterile. Explain why this is so.* This plant would be an autopolyploid (2n = 32). Four sets of chromosomes would appear at meiosis which would produce sterile gametes.

[3]

(e) *Suggest three reasons why closely related species rarely form hybrids in nature.*

(i) may be geographically separated

(ii) may flower at different times

(iii) pollen may be physiologically incompatible

[3]

[OLE]

16 Genetics

Gene   The basic unit of inheritance for a given characteristic.
Allele   Alternative forms of the same gene responsible for determining contrasting characteristics (**A** or **a**).
Locus   The position of an allele on a chromosome.
Homozygous (pure breeding)   The diploid condition where both alleles are identical (**AA** or **aa**).
Heterozygous   The diploid condition where different alleles are present (**Aa**).
Phenotype   The physical or chemical expression of a characteristic.
Genotype   The genetic expression of a characteristic in terms of alleles.
Dominant   The allele which influences the appearance of the phenotype when present in the homozygous and heterozygous condition (**A**).
Recessive   The allele which influences the appearance of the phenotype when present only in the homozygous condition (**a**).
F_1 generation   The generation produced by crossing two parental stocks, (strictly, these should both be homozygous).
F_2 generation   The generation produced by crossing two F_1 organisms.
Linkage   The existence of two or more gene loci on the same chromosome.

16.1   Mendel and His Work

Mendel carried out many years of painstaking research into the inheritance of clearly differentiated characteristics in flowers. In his 'investigations' into the inheritance of a single characteristic (**monohybrid inheritance**) such as flower position in pea plants he obtained the following results:

Parents	pure breeding axial flowers	X	pure breeding terminal flowers
F_1		all axial flowers	
F_2	651 axial flowers		207 terminal flowers
F_2 ratio	3	:	1

Fig. 16.1

From these results Mendel concluded that:

1. each parental stock possesses **two** factors determining flower position;
2. only **one** factor from each parent is present in the gamete;
3. the factor determining flower position in the axils of the leaves (axial factor) is dominant over the factor determining the terminal position for a flower (terminal flower).

186

The separation of parental factors is known as Mendel's first law, or the **principle of segregation**:

The characteristics of an organism are determined by internal factors appearing in pairs. Only **one** *of a pair of factors can be present in a single gamete.*

The ratio of dominant phenotypes to recessive phentotypes of **3 : 1** is called the **monohybrid ratio**.

In terms of modern genetics, Mendel's experiment can be written as in Fig. 16.2.

Let:

A represent axial flower (dominant)
a represent terminal flower (recessive)

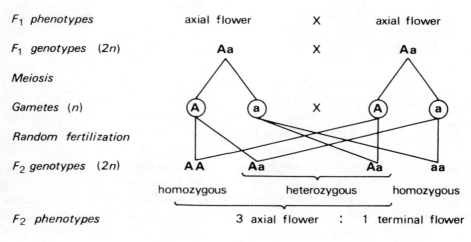

Fig. 16.2

This shows the correct way to set out a genetic diagram so as to show the various stages in the pattern of inheritance.

When Mendel investigated the inheritance of **two** pairs of contrasted characteristics (**AABB** and **aabb**) he discovered four types of offspring in the F_2 phenotypes produced from the original homozygous parents. The proportions of each phenotype approximated to **9 : 3 : 3 : 1** and is called the **dihybrid ratio**.

(*Note*: This ratio applies only to characteristics controlled by genes on different chromosomes.) On the basis of these results Mendel concluded that the two pairs of characteristics which are present in the F_1 generation separate and behave independently from one another in subsequent generations.

This forms the basis of Mendel's second law, or as it is known, the **principle of independent assortment**:

> **any one** *of a pair of characteristics may combine*
> *with* **either one** *of another pair.*

The four alleles segregate (**A**, **a**, **B** and **b**) and randomly assort (**recombination**) to give four different arrangements of alleles (four different phenotypes). Mendel's hypotheses are summarized in terms of modern genetics as follows:

1. Each characteristic of an organism is controlled by a pair of **alleles**.
2. If an organism has two different alleles for a given characteristic, one may be expressed (the **dominant allele**) to the exclusion of the other (the **recessive allele**).
3. During meiosis each pair of alleles separates (segregates) and each gamete receives one of each pair of alleles (the principle of segregation).
4. During gamete formation, either one of a pair of alleles may enter the gamete cell (random combination) with either one of another pair (the principle of independent assortment).
5. Each organism inherits one allele for each characteristic from each parent.

16.2 Testcross

In order to determine the genotype of an organism displaying a dominant phenotype the organism is crossed with a homozygous recessive organism. If **all** the offspring show the dominant phenotype the organism is homozygous dominant, but if a 1 : 1 ratio of dominant to recessive phenotypes results then the organism is heterozygous.

16.3 Linkage

Since each organism possesses thousands of characteristics and only a small number of chromosomes in each cell, it follows that each chromosome carries a large number of genes. The genes on a single chromosome form a **linkage group** and usually pass into the same gamete to be inherited together. These genes do not usually undergo independent assortment and fail to produce a 9 : 3 : 3 : 1 ratio in a dihybrid inheritance situation. Because of linkage both sets of alleles. say **A** and **B** on one pair of homologous chromosomes and **a** and **b** on another pair of homologous chromosomes, will stay linked together. Pure breeding parents will therefore produce F_1 heterozygotes and an F_2 generation showing both dominant and recessive characteristics in the ratio 3 : 1. When such a ratio is seen in dihybrid inheritance it indicates the existence of linkage.

Invariably, though, this 3 : 1 ratio is rarely seen in practice as *total* linkage is rare. It is more common to find approximately equal numbers of parental phenotypes and a smaller number of phenotypes showing new combinations of characteristics. The latter phenotypes are called **recombinants** and are the result of **crossing over**. Crossing over of alleles between homologous chromosomes occurs during **chiasmata** formation in prophase I of meiosis (see section 15.3). The frequency with which recombination occurs can be calculated using the formula:

$$\frac{\text{number of individuals showing recombination}}{\text{number of offspring}} \times 100.$$

The **recombination frequency** or crossover frequency represents a value called the **crossover value**. This value demonstrates that genes are arranged linearly along the chromosome and reflects the relative positions of genes on a chromosome. High crossover values indicate the increased likelihood that crossing-over has occurred and that the two genes are situated at some distance from each other. These values can be used in determining the positions of genes on chromosomes (**gene mapping**).

16.4 Sex Determination

The sex of an organism is determined by sex chromosomes (either **X** or **Y**) called **heterosomes**. All other chromosomes are called **autosomes**. In the majority of animals there is a pair of heterosomes. In man and many species the genotype of the male is **XY** and the female **XX**. In birds, moths and butterflies the sex genotypes are reversed. The sex with the **XX** chromosomes is called **homogametic** and the sex with **XY** chromosomes is called **heterogametic**. Each gamete contains only one sex chromosome. In most organisms the **Y** chromosome carries very few genes and therefore is often described as **genetically empty**.

16.5 Sex Linkage

Genes carried on the sex chromosomes are said to be **sex-linked**. In all sex-linked traits the gene responsible for the characteristic is carried on the **X** chromosome. Because the **X** chromosome is larger than the **Y** chromosome, characteristics controlled by genes carried on the **non-homologous portion** of the **X** chromosome appear in males even if they are recessive. Females who are heterozygous for a characteristic carried on the **X** chromosome are carriers of **sex-linked traits**. They appear phenotypically normal but half of their gametes carry the recessive gene.

Haemophilia, red–green colourblindness and premature balding are all examples of sex-linked traits.

16.6 Incomplete Dominance

It is not uncommon for two or more alleles to fail to show complete dominance or recessiveness due to the failure of any allele to be dominant in the heterozygous condition. This is called **incomplete dominance**, **co-dominance** or **blending**. It occurs in plants and animals and usually the heterozygote has a phenotype which is intermediate between the homozygous dominant and recessive conditions. The F_2 generation produced by cross-breeding the F_1 offspring of homozygous parents will have a **1 : 2 : 1** ratio of phenotypes showing a 1 : 1 ratio of parental phenotypes and twice as many intermediate forms.

16.7 Multiple Alleles

Many of the characteristics studied in genetics are controlled by a single gene which may appear in one or two forms (alleles), e.g. **A** or **a**. Some characteristics, such as blood group, are controlled by a single gene (**I**) which has more than two alleles. In this case three, represented by **A**, **B** and **O**. This is known as the multiple-allele condition. The gene locus is represented by the symbol **I** and the alleles by **A**, **B** and **O**. The blood groups corresponding to the different genotypes are as follows:

genotype	blood group (phenotype)
$I^A I^A$	A
$I^A I^B$	A
$I^B I^B$	B
$I^B I^O$	B
$I^A I^B$	AB
$I^O I^O$	O

16.8 Worked Examples

16.1

Explain the meaning of the following genetic terms:

(a) heterogametic The sex whose diploid cells do not contain two X chromosomes. In man and many mammals the male is heterogametic, carrying an X and a Y chromosome.

(b) autosomes All chromosomes other than the sex chromosomes. Autosomal chromosomes occur in pairs.

(c) lethal in the homozygous condition Certain single genes may influence mortality. The genes are recessive and in the homozygous condition will give rise to death during foetal development or later in life, e.g. an allele which produces yellow fur in the homozygous condition in mice also leads to premature death.

[6]
[AEB]

16.2

(a) Distinguish between the terms (i) phenotype and (ii) genotype.

(i) phenotype is the physical or chemical expression of a characteristic

(ii) genotype is the genetic expression of inherited characteristics

[2]

(b) If two heterozygous black guinea pigs are mated, what is the probability of their producing (i) a homozygous recessive (white) offspring and (ii) a homozygous black offspring?

(i) probability of a homozygous recessive (white) offspring

 25% (1 in 4)

(ii) probability of a homozygous black offspring

 25% (1 in 4)

[4]

(c) (i) In cats, a gene B is responsible for black fur colour and an allele L is responsible for yellow fur. These genes show incomplete dominance and the hybrid BL is tortoiseshell. The genes are also sex-linked.
   Complete the table below, to show the gametes and possible genotypes in the parents, F_1 and F_2 generations, if a black female is mated with a yellow male.

	Black female		Yellow male	
Parents	$X^B X^B$	x	$X^L Y$	
Gametes	X^B X^B	x	X^L Y	
F_1 generation	$X^B X^L$; tortoiseshell ♀	$X^B Y$ black ♂	$X^B X^L$; tortoiseshell ♀	$X^B Y$ black ♂
Gametes	X^B X^L	x	X^B Y	
F_2 generation	$X^B X^B$; black ♀	$X^B Y$ black ♂	$X^L X^B$; tortoiseshell ♀	$X^L Y$ yellow ♂

[7]

(ii) To what extent are the male and female kittens produced distinguishable by their coat colour?

Not completely distinguishable. All yellow kittens are male and tortoiseshell kittens are female, but one half of the black kittens will be female and the other half will be male.

[4]
[L]

16.3

In mice, the dominant allele **B** determines black coat colour; brown coat colour results from a recessive allele **b**. The dominant allele **N** determines hair which grows straight; wavy hair is caused by a recessive allele **n**.

A cross between two mice is shown below.

Parents	*Phenotype:*	Black, wavy-haired × Brown, straight-haired
	Genotype:	**BB nn** × **bb NN**
	Gametes:	**Bn** **bN**
F_1 *Generation*	*Phenotype:*	*All black and straight-haired*
	Genotype:	**BbNn**

(i) The parents shown are homozygous. What does this term mean?

the condition where both alleles are identical.

.... [1]

(ii) A member of the F_1 generation was used in further breeding experiments. Show the genotypes of the gametes which would have been produced by an F_1 mouse.

Gametes ...BN.....Bn.....bN.....bn............ [2]

(iii) The above F_1 generation could be produced by crossing mice whose genotypes differ from those shown.

   What would be the genotypes of the parents of this second cross?

GENOTYPES ...BBNN............. × ...BBNN............... [2]

[SEB]

16.4

Bateson and Punnett crossed sweet pea (Lathyrus odorata) plants with purple flowers and long pollen grains with plants with red flowers and round pollen grains. The F_1 plants were all purple-flowered with long pollen grains. However, in the F_2 there were

   4831 *purple, long*
    390 *purple, round*
    393 *red, long*
   1338 *red, round plants,*

a ratio of about 11 : 1 : 1 : 3.

(a) What theoretical ratio would you expect the four classes to show in the F_2?

9 purple long : 3 purple round :

3 red long : 1 red round plants

(b) How would you explain the experimental ratio?

Genes for flower colour and pollen grain length are on same chromosome, i.e. linked. This linkage is not complete however, and the genes do not undergo independent assortment.

....

(c) By what process could such new combinations as purple flowers and round pollen grains arise?

   crossing over of genes on homologous chromatids at chiasmata in prophase I of meiosis

.... [6]

[UCLES]

16.5

A male housefly homozygous for brown-coloured body and normal antennae was crossed with a female homozygous mutant fly which had a black body and forked antennae. All the offspring obtained possessed brown bodies and normal antennae. The F_1 males were mated with females having the same genetic make-up as the mother and they produced the following offspring:

>*230 flies with brown bodies and normal antennae;*
>*20 flies with brown bodies and forked antennae;*
>*18 flies with black coloured bodies and normal antennae;*
>*236 flies with black bodies and forked antennae.*

(a) Using suitable symbols to represent the alleles of the genes involved in the experiment, explain the genotypes and phenotypes of the flies in:
  (i) the parental generation;
  (ii) the F_1 generation;
  (iii) offspring of the final mating.

(b) What proportions are shown by this offspring of the final mating for
  (i) body colour alone;
  (ii) antenna shape alone?

(c) Suggest with reasons what your answers to (a)(i) and (ii) indicate about the dominance of the genes concerned.

(d) The number of progeny showing recombination expressed as a percentage of the total number of offspring obtained is termed the crossover value for the two loci involved. Determine the crossover value for body colour and antenna characteristics.

[OLE]

(a)   Let **B** = brown-coloured body
       b = black-coloured body
       N = normal antennae
       n = forked antennae

(i) *Parental generation*

BBNN	and	**bbnn**
brown body		black body
normal antennae		forked antennae

(ii) *F_1 generation*

BbNn
brown body, normal antennae

(iii) *Offspring*

BbNn	**Bbnn**	**bbNn**	**bbnn**
brown body	brown body	black body	brown body
normal antennae	forked antennae	normal antennae	forked antennae

The two phenotypes, black body, normal antennae, and brown body, forked antennae, show recombination of genes which occurred during the production of the F_1 generation.

(b) (i) *Body colour*

230 brown + 20 brown = 250 brown
236 black + 18 black = 254 black

The body colour ratio is therefore approximately 1 : 1.

193

(ii) *Antennae length*

230 normal + 20 normal = 250 normal
236 forked + 18 forked = 254 forked

The antennae length is therefore approximately 1 : 1.

(c) Black and brown and normal and forked are alleles. The parents are homo-zygous. Since the F_1 generation is brown and normal, brown is dominant to black and normal is dominant to forked.

(d) Cross-over value is $\dfrac{38}{504} \times 100 = 7.5\%$.

16.6

In Drosophila melanogaster *the gene for grey body* (y^+) *is dominant to that for yellow body* (y); *the gene for red eyes* (w^+) *is dominant to that for white eyes* (w); *the gene for long wings* (m^+) *is dominant to that for miniature wings* (m). *The genes* y^+, w^+, m^+, *are linked and are situated on the X chromosome.*

(a) *A wild type male was crossed with a yellow-bodied, white-eyed female and produced males with yellow bodies and white eyes and females with grey bodies and red eyes. When these flies were interbred the offspring consisted of 1002 flies with grey bodies and red eyes, 1004 flies with yellow bodies and white eyes, 15 flies with grey bodies and white eyes and 16 flies with yellow bodies and red eyes.*

  (i) *Show, using the symbols given, how the F_1 flies were produced and the results of their inbreeding.*
  (ii) *What is the cross-over value between grey body and red eyes?*

(b) *A wild-type male was crossed with a white-eyed miniature-winged female and produced males with white eyes and miniature wings and females with red eyes and long wings. When these flies were interbred the offspring consisted of 827 flies with red eyes and long wings, 832 flies with white eyes and miniature wings, 419 flies with red eyes and miniature wings and 459 flies with white eyes and long wings.*

  What is the cross-over value between red eyes and long wings?

(c) *If the cross-over value between grey body and long wings is 36.1 show, by means of a drawing, the relative positions of the genes for grey body, red eyes and long wings on the X chromosome.*

[OLE]

(a) (i)

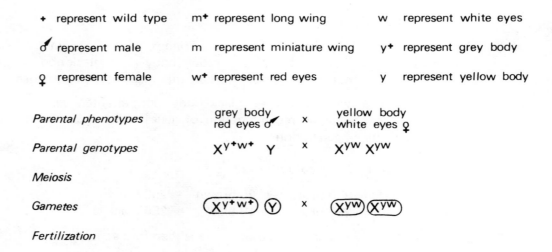

+ represent wild type	m^+ represent long wing	w represent white eyes
$\male$ represent male	m represent miniature wing	y^+ represent grey body
$\female$ represent female	w^+ represent red eyes	y represent yellow body

Parental phenotypes   grey body red eyes $\male$ x   yellow body white eyes $\female$

Parental genotypes $X^{y^+w^+}$ Y   x X^{yw} X^{yw}

Meiosis

Gametes $\boxed{X^{y^+w^+}}$ $\boxed{Y}$ x $\boxed{X^{yw}}$ $\boxed{X^{yw}}$

Fertilization

194

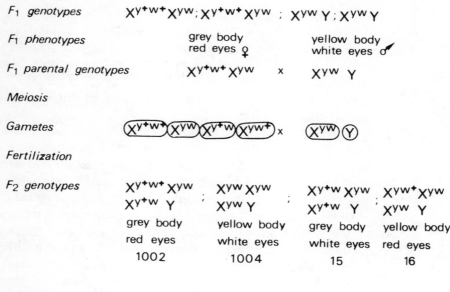

F_1 genotypes	$X^{y^+w^+}X^{yw}$; $X^{y^+w^+}X^{yw}$; $X^{yw}Y$; $X^{yw}Y$	
F_1 phenotypes	grey body red eyes ♀	yellow body white eyes ♂
F_1 parental genotypes	$X^{y^+w^+}X^{yw}$ x	$X^{yw}Y$

Meiosis

Gametes $\;\;\;(X^{y^+w^+})(X^{yw})(X^{y^+w})(X^{yw^+})$ x $(X^{yw})(Y)$

Fertilization

F_2 genotypes				
$X^{y^+w^+}X^{yw}$ $X^{y^+w^+}Y$	$X^{yw}X^{yw}$ $X^{yw}Y$	$X^{y^+w}X^{yw}$ $X^{y^+w}Y$	$X^{yw^+}X^{yw}$ $X^{yw^+}Y$
grey body red eyes 1002	yellow body white eyes 1004	grey body white eyes 15	yellow body red eyes 16

(ii) Cross-over value is $\dfrac{31}{2037} \times 100 = 1.5\%$.

(b) Cross-over value is $\dfrac{878}{2537} \times 100 = 34.6\%$.

(c)

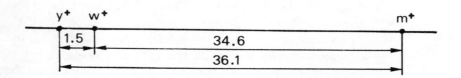

16.7

Describe the principles of genetics formulated by Mendel. Relate these principles to chromosomes and their behaviour during nuclear division and fertilization. [20]
[L]

Mendel made the first significant contribution to the study of inheritance. He proposed that characteristics, e.g. height of plant, are determined by factors that occur in pairs. In some cases both factors may be identical, e.g. tall, TT, and short, tt, or they may be different, Tt. If the two factors are unlike one may be expressed, the dominant factor (T) to the exclusion of the other, the recessive factor (t).

Each pair of factors separate so that only one factor is present in each gamete. This is known as Mendel's principle of segregation. Gametes combine randomly during fertilization and give rise to offspring whose physical appearance (phenotype) depends upon the distribution of factors in the parental generation. If both sets of parents were pure-breeding (homozygous) the offspring (F_1 generation) would all be tall, but they would also carry the recessive factor. These would be called hybrids and have the heterozygous genotype, Tt. If these heterozygous individuals cross-breed, their offspring (F_2 generation) would exhibit a ratio of 3 dominant phenotypes to 1 recessive phenotype. This ratio is called the monohybrid ratio.

In the case of the inheritance of two characteristics, height of plant and shape of seeds, Mendel observed that the two pairs of characteristics which combine in the F_1 generation separate and behave independently from one another in subsequent generations. The ability of any one of a pair of chromosomes to combine with either one of another pair forms the basis of Mendel's principle of independent assortment. If the original parental generation was homozygous the F_2 generation would produce a dihybrid ratio of 9 : 3 : 3 : 1. Mendel's principles of genetics can now be explained in terms of chromosomes and their behaviour during nuclear division and fertilization.

Diploid cells contain homologous pairs of chromosomes and each chromosome bears a number of sites called loci at which genes are situated. The form taken by each gene on each chromosome is called an allele and it may be identical, TT or tt. These are called homozygous genotypes. If the alleles are different, Tt, the genotype is called heterozygous. T,tall is dominant to t,short.

During anaphase I of meiosis the homologous chromosomes separate and one of the homologous chromosomes passes into each gamete cell. This movement accounts for the principle of segregation. The random distribution of each allele from a pair of homologous chromosomes is independent of the distribution of alleles of other pairs due to the random alignment of homologous chromosomes on the equatorial spindle during metaphase I. This movement accounts for the principle of independent assortment and leads to a variety of allele recombinations in the gamete cells.

During fertilization, which is random with respect to gamete cells, each gamete contributes one homologous chromosome to the zygote, thus restoring homologous pairs of chromosomes and bringing together pairs of alleles in new combinations. Because of the effects of dominance and recessiveness, new combinations of characteristics may appear.

The above accounts refer to characteristics controlled by genes situated on *different* chromosomes. In the case of linkage all the genes on a single chromosome usually pass into the same gamete. They do not exhibit independent assortment and fail to produce 9 : 3 : 3 : 1 dihybrid ratios. Furthermore, total linkage is rare owing to crossing over of alleles during chiasmata formation. This produces recombinants which are a major source of genetic variation.

16.8

(a) *How is sex genetically determined in birds and humans?* [4]

(b) (i) *A woman has a haemophiliac son and three normal sons. What is her genotype and that of her husband with respect to this gene? Explain your answer.* [8]

(ii) *Could she have a haemophiliac daughter? Explain your answer giving your reasons.* [5]

(c) *A population of human beings will contain many more colour-blind individuals than haemophiliacs although the genes are transmitted in the same way. Explain this difference in frequency.* [3]

[L]

(a) Sex is determined by the presence of the sex chromosomes, X and Y (heterosomes). In birds the heterogametic sex is the female XY, and the homogametic sex is the male XX. In humans the heterogametic sex is the male XY, and the homogametic sex is the female XX. During gamete formation one sex chromosome passes into each gamete.

(b) (i) Haemophilia is a sex-linked characteristic carried by a recessive gene on the X chromosome. Since the father provides a Y chromosome to his sons the haemophiliac gene must come from the mother, Xh. As only one son out of four is haemophiliac the mother must be heterozygous, XXh. In this condition she is a carrier. The father could be haemophiliac too, XhY, but there is not enough information given to produce this conclusion. It is most likely that the father's genotype is XY.

(ii) If the father's genotype is XY this woman could not have a haemophiliac daughter. The genotype of the daughter would be either XX, normal, or XXh, carrier. In order to produce a haemophiliac daughter the father must be haemophiliac XhY and the mother must either be a carrier XXh or haemophiliac XhXh. In the former case 50 per cent of the female offspring could be haemophiliac and in the latter case all the female offspring would be haemophiliac. In the example given in (b)(i) half of the ova would carry the normal X chromosome and half of the ova would carry the haemophiliac-bearing chromosome Xh. The husband would only produce a haemophiliac daughter if his sperm carried the haemophiliac allele on the sex chromosome, Xh.

(c) The selection pressure against haemophilia is greater than that against colour-blindness since the former affects survival rates. Haemophiliacs are not as likely to reach sexual maturity and so the haemophiliac genes are lost from the population.

17 Evolution and Speciation

Evolution The development of complex organisms from pre-existing simpler organisms over the course of time.

Neo-Darwinism The theory of organic evolution by the natural selection of genetically determined characteristics.

Homology Two or more organs are said to be homologous if they have a similar basic structure, a similar histological appearance, a similar embryonic development and similar relative positions in different species.

Adaptive radiation The principle by which the structure of a homologous organ is differentiated to perform a variety of different functions.

Analogy Two or more structures, physiological processes or modes of life are said to be analogous if they show adaptations to perform the same functions but have no close phylogenetic links.

Natural selection The process by which those organisms with characteristics advantageous to their environment survive and reproduce at the expense of less well-adapted organisms.

Species A group of organisms capable of interbreeding and producing fertile offspring.

Population A group of organisms of the same species living within the same community.

Hardy–Weinberg equilibrium The frequency of dominant and recessive alleles in a population will remain constant from generation to generation, assuming that the population is large, mating is random, no mutations occur and that there is no emigration from, or immigration into, the population.

Mutation A change in the amount of, or structure of, DNA in an organism which leads to a change in a particulr phenotype.

17.1 Evolutionary Theory

Present-day evolutionary theory (neo-Darwinian evolution) is based on evidence from a broad range of sources and supported by a wealth of otherwise unrelated observations. Individually, much of this evidence is open to various interpretations but collectively it supports a theory of **the progressive development of complex organisms from simpler organisms**.

17.2   The Work of Charles Darwin

The concept of evolution originated long before Darwin. For almost two thousand years man has noticed the basic structural and functional similarities which exist between organisms. During Darwin's five-year trip in HMS *Beagle* he gained a great deal of knowledge concerning the diversity of living organisms. This information along with the result of his work on the selective breeding of pigeons and other domestic animals led him towards developing the concept of natural selection. This theory of natural selection was also proposed by A. R. Wallace and was jointly presented by Darwin and Wallace in July 1858. The following year Darwin published his findings more fully in his book *On the Origin of Species by Means of Natural Selection*.

17.3   Evidence for Natural Selection

The evidence for natural selection is based on the following three observations and two deductions:

(a) Observation 1

Individuals within a population have an **enormous reproductive potential**. (This observation was based on the work of the Rev. Thomas Malthus on the reproductive potential of man.) For example, the shore crab produces 4 million eggs per season.

(b) Observation 2

The **numbers** of individuals in a population **remain** approximately **constant**. Environmental factors such as the availability of food, space, light, etc. limit population sizes.

(c) Deduction 1

Within any population many **individuals fail to survive or reproduce**. Thus there is a struggle for existence. Competition for environmental resources between organisms of the *same* species (**intraspecific competition**) and between members of *different* species (**interspecific competition**) influences reproductive and survival rates.

(d) Observation 3

Variation exists between organisms within all populations.

(e) Deduction 2

Under the intense competition of numbers of organisms in a population, any variations which favour survival of an organism in a particular environment will increase the likelihood of it reproducing and leaving offspring (this is called a selective advantage). Conversely, organisms possessing less favourable variations will be at a disadvantage and their chances of reproducing and leaving offspring will be decreased (selective disadvantage).

17.4 Evidence for Evolution

(a) Palaeontology

Fossils are the preserved remains of living organisms. They may be of whole organisms, parts of organisms or imprints and are found in most sedimentary rocks. The oldest strata of rocks contain fossils with simple structures. Fossils in later strata appear to be related to the earlier simpler forms but are more complex structurally, e.g. molluscs and crustaceans. Many of the species found in the early strata are not found in later strata. This is interpreted in terms of the times at which the species originated and became extinct. There are 'gaps' in the fossil record for the majority of groups of organisms. However, the fossil record of the horse, from the early Eocene to the present, is a good example of progressive adaptation to changing environments.

(b) Geographical Distribution

Plant and animal species are discontinuously distributed throughout the world. The *same* species does not occupy similar habitats in similar geographical areas as might be expected, e.g. lions are found in the Savannah of Africa and not in the Pampas of South America. However, closely related specimens may be found in these areas, e.g. cougars in the Pampas. Observations such as these are explained in terms of an ancestor originating in a particular area and dispersing outwards from that point to adjacent land masses. Geological evidence suggests that the relative positions of these land masses then changed with time. Once dispersed, each sub-group of the species gradually adapted according to the selection pressures imposed by the different environmental conditions found on these land masses, e.g. the common ancestor of the camel in Africa and the llama of South America is thought to have originated in North America.

Likewise, the unique presence of marsupial mammals in Australasia is explained in terms of the separation of this continent from the other land masses prior to the appearance of placental mammals. Marsupials continued to thrive in Australasia free from the competition of the more advanced placentals which eliminated marsupials elsewhere. In Australasia, adaptive radiation occurred and marsupials now occupy the same ecological niches as placental mammals elsewhere, e.g. kangaroo and antelope, koala and sloth, wombat and prairie dog, marsupial mouse and mouse. This phenomenon is called convergent evolution and suggests the existence of an evolutionary mechanism. The distribution of finches on the Galapagos Islands is further evidence that such a mechanism exists.

(c) Classification

The hierarchical classificatory system of Linnaeus consists of the following groups: phylum, class, order, family, genus and species: it is based on the structural similarities that exist between organisms. These similarities suggest the existence of an evolutionary relationship between organisms. In itself, this evidence is not conclusive evidence of such a process.

(d) Plant and Animal Breeding

Man has demonstrated that it is possible, by selective breeding, to produce desirable characteristics in offspring. 'New' strains of plants which are resistant to diseases, germinate rapidly, bear large fruit etc., and new breeds of animals, e.g. cows with high milk yield and resistance to certain diseases, have been developed. If 'artificial' selection can bring together such characteristics in a short space of time it is argued that 'natural' selection could do so too, given the time-scale of millions of years.

(e) Comparative Anatomy

Organisms which show striking similarities, e.g. lion and dog, are assumed to have a close relationship. Each of their constituent organs has a similar embryological development, structure and function, e.g. the **pentadactyl limb**. Such organs are described as **homologous**. These structures are assumed to have had a common ancestry. The fact that a seal's flipper is used for swimming, a bat's wing for flying and a horse's leg for running suggests that these are limb adaptations to different environmental conditions and modes of life, a phenomenon called **adaptive radiation**. All three organisms are believed to have had a common ancestor.

Many organs carry out similar functions, e.g. the eyes of squid and man and the wings of birds and moths. They have similar structures but different embryological origins. They are described as **analogous** and their similar structure and function is explained in terms of **convergent evolution**. This is the development of similar structures by gradual adaptation to the environment in order to fulfil a common function. It suggests the existence of an evolutionary process and mechanism. The existence of **vestigial organs**, e.g. the appendix and coccyx of man and of herbivorous mammals, implies the existence of a common ancestor and also suggests an evolutionary process.

(f) Comparative Embryology

The embryological development of all vertebrate groups shows striking structural similarities, particularly during cleavage, gastrulation and organogeny. Furthermore, the development stages seen in, for example, a mammal, show many of the structures seen in the adult stages of other, more primitive chordate animals, e.g. visceral clefts, segmental myotomes and a single circulation with a two-chambered heart. This is interpreted in terms that all chordate animals are related, by descent, from a common ancestry.

(g) Comparative Biochemistry

Similar molecules are found in a complete range of organisms. For example, cytochrome C, a conjugated protein used in respiratory pathways, is found in the majority of organisms. It contains a polypeptide chain whose amino acid sequence shows striking similarities in all of these organisms. Furthermore, the similarities are greater within a group than between groups, e.g. all primates have a similar sequence which is more similar to other mammals than to birds etc. There are many more examples which suggest a form of biochemical homology.

17.5 Modern Views on Evolution

The theory of evolution has been amplified in the light of more recent developments in molecular biology, genetics, ecology and behaviour. The theory, now described as **neo-Darwinism**, is based on the natural selection of genetically determined characteristics.

17.6 Population Genetics

The characteristics of organisms are determined by genes acting independently (**discontinuous variation**), or in conjunction with the environment (**continuous variation**) to determine phenotypic characteristics and variation (indicated). Phenotypes adapted to the environment survive and reproduce; phenotypes not adapted to the environment do not survive and reproduce. Thus evolution is now seen in terms of genes being 'selected for' or 'selected against'.

All the alleles and genes in a sexually reproducing population form the **gene pool**. If the gene pool changes from generation to generation the population is undergoing evolutionary change. It is possible to calculate the *frequency* of alleles in a population and use this to determine the *extent* of evolutionary change and to *predict* the form of this change.

The Hardy–Weinberg equation is used to calculate allele and genotype frequencies.

If p = dominant allele frequency, and
q = recessive allele frequency, then

$$p + q = 1.$$

Substituting a dominant allele, **A**, for p and a recessive allele, **a**, for q it can be seen that the frequency of genotypes can be determined.

p^2 (**AA**)   = homozygous dominant genotype;
$2pq$ (**2Aa**) = heterozygous genotype;
q^2 (**aa**)   = homozygous recessive genotype.

Usually the frequency of the homozygous recessive genotype (q^2) is stated and the other genotype and allele frequencies can be determined by substitution, using the equation:

$$p^2 + 2pq + q^2 = 1.$$

17.7 Selection

The environment is the agent of natural selection. It 'selects' which organisms will survive and reproduce by acting on the various phenotypes within the population. As the population increases in size, certain environmental factors become limiting, e.g. food availability in animals and light in plants, and these produce a **selection pressure** which can vary in intensity. As a result of this selection certain alleles are passed on to the next generation by virtue of the differential advantages they exhibit when expressed as phenotypes.

One source of selection is man. By breeding closely related organisms, a technique called **inbreeding**, desirable characteristics can be intensified. This was used to produce high yields of milk, wool and eggs but it leads to a reduction in fertility and has now been largely replaced by **outbreeding**. In this case breeding takes place between members of different varieties or strains in animals. In plants breeding occurs between closely related **species**. In both cases the progeny are known as **hybrids** and these have phenotypes with characteristics superior to either of the parental stocks. This **hybrid vigour** results from the increased heterozygosity produced by gene mixing.

17.8   Natural Selection

The major source of selection is the environment working on variation introduced into a population by sexual recombination during meiosis and by random mutations. A mutation may affect the structure of a single gene or alter the structure and number of chromosomes in a cell. The majority of **gene mutations** occur during nuclear division. Specific environmental conditions such as radioactivity or other forms of ionizing radiation or mutagenic chemicals can increase the frequency of these mutations. Structural changes in chromosomes alter the sequence of genes, e.g. translocation. Numerical changes in chromosomes are rarer, e.g. **polyploidy** which produces multiple sets of chromosomes as in triploid and tetraploid plants. Alternatively, individual chromosomes can be duplicated $(2n + 1)$, e.g. 47 chromosomes in Down's syndrome, or deleted $(2n - 1)$.

There are many examples of natural selection which can be seen in action today. In the majority of cases changes in allele frequency have resulted from increases in selection pressure. Examples include **industrial melanism** in moths, **polymorphism** in snails, **metal tolerance** in plants, antibiotic resistance in bacteria, pesticide resistance, e.g. warfarin in rats and DDT in mosquitoes, and sickle-cell anaemia.

17.9   Speciation

In the majority of cases speciation occurs by a single species giving rise to a new species. This is called **intraspecific speciation**. Several factors may influence the process but in all cases **gene flow** within the population must be interrupted. Each subgroup becomes **genetically isolated** and allele and genotype frequencies change owing to natural selection acting on a range of phenotypes produced by sexual recombination and random mutation. If the races and subspecies produced continue to remain genetically isolated over a prolonged period of time they will be unable to interbreed if brought together again. They will then be considered to be separate species. In all cases some form of **isolating mechanism**, e.g. geographical, seasonal, ecological, behavioural, mechanical, isolation, must exist.

If speciation occurs, while the populations are separated it is called **allopatric speciation**. The usual cause of the separation is some form of geographical barrier, e.g. the sea or a mountain range. Speciation of the finches on the Galapagos Islands is thought to have occurred in this way.

If speciation occurs while the populations are occupying the same geographical area it is called **sympatric speciation**. Usually, sympatric speciation follows a period during which some form of isolating mechanism has existed, e.g. the hooded crow and the carrion crow in the British Isles are undergoing this form of speciation. This form of speciation is far less common than allopatric speciation.

17.10 Worked Examples

17.1

(a) *What is meant by 'organic evolution'?*

The development of complex organisms from pre-existing simpler organisms over the course of time.

[3]

(b) *Briefly describe the use of the following for dating fossils and in each case comment on its reliability.*

(i) *evidence from sedimentary rocks* Sedimentary rocks are stratified. Younger rocks lie on top of older rocks. Each stratum has fossils associated with a particular age. Not very reliable.

(ii) *radioactive decay* Radioactive elements decay at a constant rate. By comparing the ^{14}C in dead organisms with that in living organisms the age of the organism can be estimated. Relatively reliable.

[4]

(c) *Briefly describe how each of the following can be used to provide evidence for evolution.*

(i) *homologous organs*

If an organ has a similar basic structure, a similar histological appearance, a similar embryonic development and similar relative positions in a range of organisms it suggests they are related by evolution.

(ii) *development of vertebrate embryos*

All vertebrate embryos show striking structural similarities. Furthermore the developmental stages in mammals show many of the structures seen in adult stages of other, more primitive chordate animals.

(iii) *industrial melanism*

The unequal distribution of dark and light forms of Biston betularia in soot-polluted and non-polluted areas is a result of selective predation by birds. This is an example of natural selection in action.

[9]

[L]

17.2

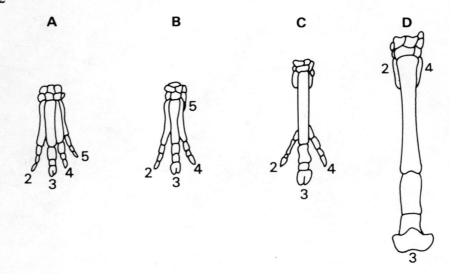

A   **B**   **C**   **D**

Fig. 17.1

Figure 17.1 shows, in chronological sequence from A to D, fossils of the forelimb skeletons of four related mammals.

(a) To which modern mammal are these most closely related?

. . . . *horse* . . . . [1]

(b) Describe briefly the structural changes seen in the fossil sequence.

. . . . *loss of 5th digit* . . . .

. . . . *reduction of 2nd and 4th digits* . . . .

. . . . *increase in length and prominence of 3rd digit* . . . .

. . . . *appearance of a true 'hoof'* . . . . [4]

(c) What were the possible adaptive advantages of these structural changes?

. . . . *increased leverage for running* . . . .

. . . . *raised the body off the ground* . . . .

. . . . *increased the speed of running* . . . .

. . . . *increased the strength of the limb* . . . . [4]

(d) State two methods by which these fossils might be dated.

first method . . . . *radioactive carbon dating of fossil* . . . .

. . . .

second method . . . . *dating the age of the rocks in which the* . . . .

fossils are found . . . . [2]

[L]

17.3

Parts A to D in Fig. 17.2 show the heads of four of the ten species of finch inhabiting a volcanic island in the Galapagos group, some 600 miles from the mainland of Ecuador, South America.

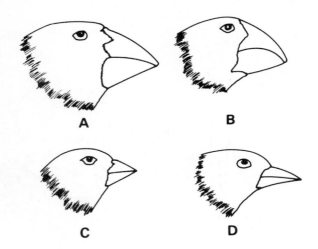

Fig. 17.2

(a) *What major difference between the four species is shown in the diagrams?*

    variation in size and shape of beak .... [1]

(b) *How might the difference you have given in (a) be related to the way of life of the finches?*

    The birds feed on different foods. B probably eats seeds and D probably eats insects.

    .... [2]

(c) *These finches probably descended from a common ancestral stock of finches. Suggest how these ancestors may have reached the island.*

    They may have been blown from Ecuador to the Galapagos Islands in a gale or carried on floating logs.

    .... [2]

(d) *What evidence to support theories of how evolution occurs may be deduced from the information provided by these finches?*

    Geographical isolation of the finches led to allopatric speciation. Differences in beak shape show that adaptive radiation occurred. Each beak is adapted to a particular ecological niche. [3]

(e) *There are more species of finch than other birds on the island. Suggest an explanation for this.*

    Finches may have colonized the island before other birds and become well established. Other birds are not able to compete with them. [2]

(f) *Why is the diversity of finch types on the South American mainland less than on the island?*

    More food may be available in South America therefore selection pressure will be less. Also there are fewer isolating mechanisms on the mainland.

    [2]

(g) *The different finch species on the island do not interbreed. What might this indicate?*

They are true species and have undergone complete speciation.

They are now reproductively incompatible.

[2]

(h) *All the species of finch on the island are dull-coloured. Give two ways in which recognition might occur between sexes of the same species of finch.*

(i) bird song

(ii) behaviour pattern

[2]

[L]

17.4

How might each of the following be used as support for the theory of evolution?

(a) *Human beings possess an appendix which seems to have no function.*

'Older' mammals, as revealed by fossilized strata, have a functional appendix. This is used in cellulose digestion. The human appendix and herbivore appendix are homologous.

[3]

(b) *Marine sharks retain urea in their blood, thus maintaining their osmotic pressure close to that of the surrounding sea water. Fresh-water sharks show the same phenomenon.*

Urea in blood is an adaptation to a hypertonic environment. Fresh-water sharks must have evolved at a later stage from marine sharks but are not yet adapted to a hypotonic environment.

[3]

(c) *Viral DNA has the same basic structure as human DNA.*

Suggests that all DNA had a common origin and ancestry. Self-replicating DNA, once originated, survived selection pressures.

[2]

[L]

17.5

Discuss the genetical basis and evolutionary significance of the following:
(a) formation of new species by allopolyploidy;
(b) heavy-metal tolerance;
(c) heterostyly in Primula *sp.;*
(d) shell colour and banding in Cepaea nemoralis;
(e) sickle-cell anaemia. [OLE]

(a) Although many hybrid plants are viable they are unable to reproduce sexually because their chromosomes are unable to pair at meiosis. If the chromosome number of such a plant doubles it becomes an allopolyploid species. Its chromosomes can now pair during meiosis because each chromosome has a homologue. This plant is a new species which can reproduce sexually with other similar allopolyploid species. For example, the fertile allotetraploid *Triticale* ($2n = 56$) was formed from a cross between *Triticum* (wheat) ($2n = 42$) and *Secale* (rye) ($2n = 14$). The intermediate hybrid was sterile but when it doubled its chromosome number it formed *Triticale*.

(b) Heavy-metal tolerance is seen in the grass *Agrostis tennuis*. Some plants will grow on soils contaminated with copper, or lead or zinc, where they show resistance to these otherwise poisonous metals. The tolerance is due to the presence of polygenes, which are groups of genes with slight individual effects but important cumulative effects. They determine characteristics which exhibit continuous variation. Slight changes in the structure of a polygene can confer variation in phenotype such as tolerance to metal poisons. The heavy metals in the soil act as agents of natural selection and confer a selective advantage on certain tolerant plants. In the absence of competition from other plants these plants will thrive.

(c) Two polymorphic species of *Primula* exist, one with long anthers and a short style, called thrum-eyed, and another with short anthers and a long style, called pin-eyed. This is an example of heterostyly. The difference between these two polymorphs is due to a set of closely linked genes called a super-gene. Crossing over between these genes is rare. However, such a crossing over has given rise to a polymorph with both long anthers and a long style. It is self-fertile and has a selective advantage in areas where insects are scarce since it is capable of self-pollination.

(d) In *Cepaea nemoralis* brown shell is dominant to pink shell which in turn is dominant to yellow shell. Unbanded shells are dominant to banded shells. The genes for shell colour and banding are linked to form a super-gene. These genes determine characteristics which have such a selective advantage that they are maintained within the population. *Cepaea* is predated upon by thrushes which select their prey by visual selection. Yellow shells (appear green with the animal inside them) are camouflaged against a grass background and pink and brown are at a selective advantage on leaf litter in woods. Like-wise, banded is at an advantage over unbanded on a uniform background. Over the space of a year no shell colour or banding pattern has an advantage and the numbers of all types in the population remain fairly constant. There is also evidence which suggests that for some physiological reason heterozygotes have an advantage over homozygotes.

(e) Sickle-cell anaemia is due to the presence of S haemoglobin in the blood. A single gene mutation causes the erythrocytes in sufferers to be curved and distorted and easily destroyed. Homozygotes usually die as a result of severe anaemia whereas heterozygotes survive. The sickle-cell gene maintains a high frequency in certain populations living in regions where malaria is endemic because it confers a selective advantage. *Plasmodium* fails to complete its life cycle in damaged red cells. As malaria continues to be eradicated the gene frequency of the sickle gene will decrease.

17.6

(a) *What is meant by the terms (i) continuous variation, and (ii) discontinuous variation? Illustrate your answer with examples in each case.* [3, 3]

*(b) Explain, with reference to a **named** example in each case, how these variations arise.*

[UCLES]

(a) (i) Continuous variation refers to the situation where a single phenotypic characteristic shows a complete gradation from one extreme to another without any break. For example, the mass, linear dimension, shape and colour of organisms in a population show continuous variation; see Fig. 17.3(b).

   (ii) Discontinuous variation refers to the situation where a single phenotypic characteristic, e.g. blood group, appears in a limited number of distinct forms; see Fig. 17.3(a). In humans, there are four major blood-groups, A, AB, B and O. No intermediate forms exist.

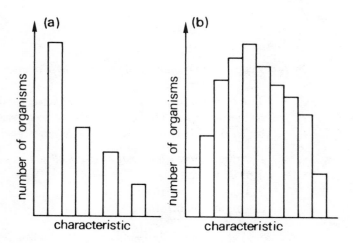

Fig. 17.3

(b) (i) The continuous variation for a factor such as mass of an elephant is produced by the combined effects of many genes forming a special gene complex, called a polygenic system. While the individual effect of each of these genes would have no effect upon the phenotype, their combined effects are significant. These effects produce a range of potential masses for the elephant. Coupled to these genetic effects are environmental influences. Influences such as food supply, temperature and health also determine phenotypic appearance and add to the extent of the continuous variation. Further variation is introduced by the genetic effects of reciprocal crossing-over of genes between chromatids of homologous chromosomes during meiosis and the orientation of these chromatids on the equatorial spindle during metaphase. Finally, a further source of variation is introduced by the random nature of the fusion of male and female gametes.

   (ii) The discontinuous variation of a phenotypic characteristic such as blood-group is partly produced by the effects of genetic segregation, genetic reassortment, random fusion of gametes as described above, and the influence of dominance. For example, in human blood-groups the allele for A is dominant to the corresponding allele for blood-group O. In many cases of discontinuous variation as illustrated by melanic and light forms of *Biston betularia*, one or two major genes, each having two or more allelic forms, may influence phenotype expression.

The major source of discontinuous variation results from either gene or chromosome mutations. An example of the former is the occurrence of sickle-cell anaemia. Here a base substitution has given rise to an abnormal form of haemoglobin (S). Chromosome mutations involve changes in the number or structure of chromosomes. This may produce changes in single chromosomes as in aneuploidy, leading to, say, Down's syndrome ($2n = 47$), or euploidy, producing either hybrid vigour or hybrid sterility in plants according to the nature of the change.

Characteristics showing discontinuous variations are independent of environmental conditions.

17.7

(a) Explain the meaning of the term allele frequency (often referred to as 'gene frequency').
[2]
The allele frequency is the rate at which a given allele passes from generation to generation. The allele will be expressed in the phenotype if it is dominant or, if it is recessive, only when present in the homozygous state.

*(b) List **three** forces which may alter the allele frequency in a small population.* [3]
1. mutations
2. non-random mating
3. sexual selection

(c) (i) The algebraic expression of the Hardy–Weinberg principle is given below:

$$p^2 + 2pq + q^2 = 1$$

where p and q are the frequencies of two alleles. State, in words, the Hardy–Weinberg principle. [3]
*(ii) 'Woolly hair' is common among Norwegian families: the hair is tightly kinked and very brittle. The allele for woolly hair (**H**) is dominant over that for normal hair (**h**). The alleles for **H** and **h** have frequencies p and q respectively. In a certain population of 1200 people, 1092 individuals have woolly hair.*
    *Assuming that the Hardy–Weinberg principle applied, calculate the frequency of occurrence of each of the genotypes **HH**, **Hh** and **hh**.*

Show all your working clearly. [8]

(i) See definition on page 198.

(ii) Since $p^2 + 2pq + q^2 = 1$,
    let p^2 = **HH** (homozygous dominant genotype),
          $2pq$ = **Hh** (heterozygous genotype),
          q^2 = **hh** (homozygous recessive genotype).

If 1092 individuals have woolly hair out of a population of 1200, 108 individuals must have normal hair (**hh**).

Therefore, the **hh** genotype frequency (q^2) = $\dfrac{108}{1200}$ = 0.09.

If q^2 = 0.09,
  q = $\sqrt{0.09}$
     = 0.3.

Since $p + q = 1$

$$p = 1 - q$$
$$= 1 - 0.3$$
$$= 0.7$$
$$\therefore p^2 = \underline{0.49} = HH.$$
$$\therefore 2pq = 2 \times (0.7 \times 0.3)$$
$$= \underline{0.42} = Hh.$$

Thus,

HH = 49%

Hh = 42%

hh = 9%

[AEB]

17.8

Discuss the mechanisms by which a new species may originate and a species become extinct.

[10]

The first part of this question could be answered by including the information given in section 17.9 on speciation.

A species may become extinct because it ceases to be adapted to its environment. In such a situation the species is at a selective disadvantage and its numbers gradually decrease as the selection pressure increases. Selection pressure can take a variety of forms. It may involve intense predation as in the case of the overexploitation of the species by man. Many species of whales are currently on the verge of extinction for this reason. Pathogenic micro-organisms have a high rate of mutation and failure on the part of a species to develop resistance quickly enough may lead to the extinction of the species.

The lack of a basic nutrient or the elimination of a link in a food chain, too, can have serious effects on the viability of a species. For example, dinosaurs are believed to have become extinct at the end of the Mesozoic because of the lack of lush vegetation on which to feed. As herbivorous dinosaurs failed to obtain enough food they died out, as did the carnivorous dinosaurs which fed on them further along the food chain.

Many species are thought to have become extinct through intense competition from a newly developed, closely related species. For example, marsupial mammals became extinct everywhere owing to intense competition from placental mammals except Australasia where there were no placental mammals.

As the population size declines, gene flow within the species is interrupted until finally the species becomes extinct.

18 Ecology

Biosphere The surface of the earth where life exists. It is irregular in thickness and in density.

Biome A region possessing characteristic physical conditions which supports animals and plants which show adaptations to these conditions.

Biomass The mass of living material, usually as dry mass per unit volume or area.

Environment The external surroundings in which an organism lives.

Ecosystem A natural unit consisting of biotic and abiotic components interacting to produce a stable system.

Community Populations of animals and plants living together in a common environment. The individuals and populations of a community interact with each other and with their non-living surroundings.

Population A group of organisms of the same species living within the same community.

Habitat The place in an ecosystem where a particular organism lives.

Ecological niche The role and status of an organism within its community. The term is also used when referring to the specific mode of life of an organism within its habitat.

Ecology is the study of the relationships of living organisms to each other and their surroundings. **Autecology** is the study of the relationships of one organism or population with its environment, while **synecology** is the study of the relationships of communities with the environment.

18.1   The Ecosystem

The ecosystem consists of a **biotic** community of living organisms interacting with each other and the non-living or **abiotic** components of the environment. The living organisms are autotrophs or heterotrophs. Autotrophs fix solar energy and manufacture complex organic food molecules, thereby providing the means by which light energy enters the ecosystem. This energy then flows through the ecosystem and is governed by the laws of thermodynamics which state:

1. Energy may be transformed from one form to another (e.g. light into potential energy of food) but cannot be created nor destroyed.
2. No process that involves energy transformation is 100 per cent efficient and some energy will always escape as heat.

Ultimately all energy that enters an ecosystem is lost from it as heat.

Inorganic materials vital for the well-being of ecosystem communities are obtained from the abiotic environment. **Macronutrients** are required in relatively large quantities and **micronutrients** in only minute amounts. These materials circulate through the ecosystem, passing back and forth between the organisms and their environment as biogeochemical cycles.

18.2 Biotic Components of an Ecosystem

(a) Food Chains and Trophic Levels

Only 1 to 2 per cent of the total available solar energy enters the food chains of living organisms. The transfer of this energy as food in plants through a sequence of organisms which eat it and then are in turn eaten themselves, is called a **food chain**. Each feeding level is called a **trophic level**.

Green plants (autotrophs or primary producers) occupy the first trophic level. They harness solar energy and incorporate it into organic molecules. These provide the food for organisms occupying the second trophic level. These are herbivores (or primary consumers). Carnivores (secondary consumers) in turn eat the herbivores, and larger carnivores eat smaller ones and so on.

The quantity of living material in the different trophic levels is called the **standing crop**, a term that may be expressed as number per unit area or as biomass.

In a stable ecosystem many food chains interconnect to form a **food web**. This shows that there is a wide range of plants for herbivores to feed on and of prey for carnivores to eat. The communities are constantly involved in complex and dynamic interactions. However, they also display a high degree of stability brought about by homeostasis. Any tendency to deviate from the norm is corrected by mechanisms of positive or negative feedback.

18.3 Ecological Pyramids

These provide a quantitative means of studying the relationships between organisms in a given ecosystem.

(a) Pyramid of Numbers

This enables us to investigate the numerical relationships between producers and consumers within an ecosystem. Organisms are placed in their trophic levels and counted. Autotrophs vary in size and numbers but there is a decrease in numbers of animals from the herbivore level towards the higher carnivore feeding levels (Fig. 18.1).

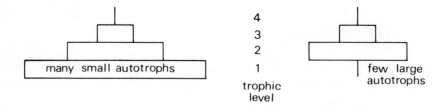

Fig. 18.1

(b) Pyramid of Biomass (Dry Mass)

The total biomass for each trophic level as calculated at the time of sampling is called the standing crop biomass of that trophic level. As energy is consumed by the organisms at each trophic level, less energy is available to organisms further along the food chain which consequently supports a smaller biomass (Fig. 18.2).

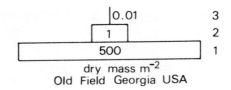

Fig. 18.2

(c) Pyramid of Energy

This takes into account the amount of energy per unit or volume that flows through each trophic level in a specific time (i.e. rate of production) (see Fig. 18.3).

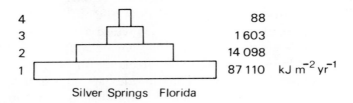

Fig. 18.3

18.4 Production Ecology

This is the study of energy flow through ecosystems. It enables us to analyse the efficiency of our various agricultural systems. The **gross primary production (GPP)** represents the rate at which solar energy enters and is fixed in autotrophs. It ultimately determines how much life the ecosystem can support. 20 per cent of the GPP is respired by the autotrophs, while the remainder, called the **net primary productivity (NPP)**, is available for growth and/or transfer to the next trophic level. Efficiency of energy transfer from autotroph to herbivore is approximately 10 per cent and from animal to animal about 20 per cent. The energy lost from each trophic level of a food chain via excreta, egesta or death passes into detritivore or decomposer food chains. The remaining energy used by heterotrophs for growth, repair and reproduction is called secondary production.

18.5 Ecological Succession

The composition and nature of a community is established over a period of time by the process of ecological succession. The abiotic characteristics of an environment determine which plants and animals will ultimately inhabit it. As an ecosystem matures with each succession (**seral stage**) the number of species in it, and its biomass, increase, and food webs become more complex. Predator–prey, parasitic and symbiotic relationships increasingly occur and there is a gradual occupation of all the ecological niches of the environment. Competition increases and may lead to specialization within organisms. This can affect the evolution of the species involved. The final or **climax community** existing once succession is complete is in dynamic equilibrium with the environment. The process of complete succession is called a **sere** and the resulting climax community may have one dominant or several co-dominant species within it.

Primary succession is the invasion and occupation of a region not previously inhabited by vegetation. Pioneer autotrophs arrive only to be replaced in due course by a second type of vegetation, which itself may be replaced, and so on. Animal communities arrive only after the vegetation is established. They also exhibit succession but their species are influenced by the types of plant present in the environment. Each temporary community changes the environmental conditions to some extent (e.g. temperature, light, humidity) which provides favourable conditions for the next community to succeed it. Secondary succession takes place in a region previously occupied by vegetation.

18.6   Zonation

This is the spatial distribution of species within a community and is brought about by varying physical conditions which dictate where each species is able to live.

18.7   Communities and Populations

The numbers and kinds of organisms within a community are determined by the interactions that take place among the populations of that community. The status of an organism within its community is called its **ecological niche**. The niche involves all of the environmental resources used by the organism (e.g. type of food consumed, where it lives, timing of activities).

18.8   Population Dynamics

This is the study of how populations grow, are maintained and decline in response to environmental fluctuations. The size of a population is dictated by three major forces:

1. **Natality**: the rate at which new individuals are added to the population by reproduction.
2. **Mortality**: the rate at which individuals are lost by death.
3. **Immigration and emigration**.

Environmental resistance is the collective name given to those factors which actively limit the growth of a population. Environmental resources and the diversity of individuals play an important part in population dynamics. The rate of growth of some populations decreases as the density increases, that is they are self-limiting or **density-independent**. The density of this type of population tends to level out before saturation point is reached. Intraspecific competition is important in density-dependent populations, competition for food, shelter and territory being several effective means of controlling population growth. Interspecific competition between different species may also occur. Such competition for resources between species limits the number of species that can exist in a particular community.

If two competing organisms occupy the same ecological niche, one usually declines in number and may become extinct. The Russian biologist Gause called this the **competitive-exclusion principle**.

Chemical competition exists among micro-organisms. Where the metabolic by-products of one microbe affect the growth of others the result is called **allelopathy**, e.g. *Penicillium* exerts an antibiotic effect against Gram-positive bacteria.

Interspecific competition also occurs between trophic levels, among predator–prey and host–parasite associations.

Some populations are not self-limited. Their growth is said to be **density-independent**. It occurs exponentially until checked by factors outside the population (e.g. changes in the biotic or abiotic environment), when there is a sudden crash. These populations suffer frequent severe oscillations in density. Often density-dependent and density-independent factors interact to regulate the size of a population.

The **maximum sustainable yield (MSY)** of a population is the rate at which individuals can be removed from a population without affecting its future productivity. This important concept is used in animal management, e.g. commercial fishing. Here only the older larger fish are netted, leaving the smaller sexually immature fish free to breed at a later stage and capitalize on the increased level of food now made available by the removal of the older fish. This process is called **rational cropping**.

18.9 Biogeochemical Cycles

These link the biotic and abiotic components of an ecosystem. Whereas energy flows in a single direction through an ecosystem, inorganic nutrients are continuously cycled and recycled round the ecosystem. Living organisms need carbon, hydrogen, nitrogen, oxygen, phosphorus and sulphur in relatively large amounts, and many other elements in much smaller quantities. They are absorbed into the bodies of the producers, passed on to the consumers through the food chain and eventually returned to the abiotic environment as egesta, excreta or dead bodies. Soil-dwelling decomposers break down these materials into a form which is readily available for the producers to use again. While these inorganic materials are required for normal growth and reproduction, excessive amounts may be toxic or inhibitory. If a specific inorganic element determines whether or not an organism can grow successfully in a particular environment, this is called a **limiting factor**.

18.10 Abiotic Components of the Ecosystem

These are subdivided into edaphic, climatic and topographic factors.

(a) Edaphic Factors

The soil provides an important link between the biotic and abiotic components of terrestrial ecosystems. It is composed of a mineral skeleton (50 to 60 per cent), organic matter (10 per cent), air (15 to 25 per cent) and water (25 to 35 per cent). It also harbours the microorganisms responsible for decomposition and recycling of materials from dead organisms to autotrophs.

(b) Climatic Factors

(i) *Light*

This is a source of energy for photosynthesis. The wavelength of light and photoperiod can have variable effects, as can the angle of incidence of the sun's rays. This varies with latitude, season, time of day and aspect of slope.

(ii) *Temperature*

The sun's radiation provides the main source of heat. Temperature is also dependent on latitude, season, time of day and aspect of slope.

(iii) *Moisture/salinity*

The hydrological cycle governs water availability on land. How much water there is in a soil depends on its water retention properties and the balance between precipitation and the effects of evaporation and transpiration. Aquatic organisms have problems of water regulation whether they inhabit fresh or salt water.

(iv) *Atmosphere*

A number of the atmosphere's gaseous components are linked to biogeochemical cycles. Wind affects growth and transpiration rates in plants and aids the dispersal of fruits and seeds. The atmosphere provides little physical support for terrestrial organisms and thus directly influences their structure.

(c) Topographic Factors

(i) *Altitude*

At high altitudes, atmospheric pressure is reduced, consequently reducing oxygen availability and increasing the rate of transpiration in plants. The temperature is lower and there are stronger winds.

(ii) *Aspect*

In the northern hemisphere, south-facing slopes receive more sunlight and are subject to higher temperatures. The reverse is the case in the southern hemisphere. The inclination of slopes is important, for the steeper they are the faster is the drainage and water run-off, making the slopes drier.

18.11 Worked Examples

18.1

    (a) Explain what is meant by an ecosystem. [5]
    (b) Explain the meaning of the following, giving examples all chosen from one named ecosystem.
       (i) pyramid of biomass
       (ii) ecological niche
       (iii) trophic levels
       (iv) food web

[15]
[L]

(a) An ecosystem is a natural unit of ecology consisting of a community of organisms (biotic component) and the non-living environmental factors (abiotic component) which influence the community. Implicit in such a definition are the concepts of energy flow through trophic levels and the cycling of matter between the biotic and abiotic components.

(b) *Rocky Seashore Ecosystem*
  (i) *Pyramid of Biomass*
      A pyramid of biomass shows the dry mass of organic matter at each
      trophic level of a food chain. The biomass of each succeeding trophic
      level is less than the one preceeding it because of energy losses along the
      food chain. For example, if a shore crab, *Carcinus*, feeds exclusively on
      periwinkles, *Littorina*, which feed, in turn, exclusively on seaweed,
      Fucus, the standing crop biomass at each trophic level would produce a
      pyramid shape if biomass was plotted, as shown in Fig. 18.3.
  (ii) *Ecological Niche*
      Each species occupies a distinct ecological niche within a given habitat.
      The niche describes not only the exact place where the species lives but
      also its role and status within the community, i.e. how it interacts with
      all other organisms in the habitat. For example, the lugworm *Arenicola*
      lives in sandy or muddy patches on rocky shores. It lives in tubes, feeds
      on detritus and is, in turn, predated upon by fish, crabs and sea birds.
  (iii) *Trophic Levels*
      A trophic level is each stage in a food chain. Each trophic level is rep-
      resented by organisms which share a particular form of nutrition as
      shown in Fig. 18.4.

Fig. 18.4

  (iv) *Food Web*
      A food web shows the feeding interrelationships which exist between the
      various food chains found within an ecosystem. It is produced by con-
      necting the food chains together and gives a more realistic description of
      how top carnivores obtain their food (Fig. 18.5).

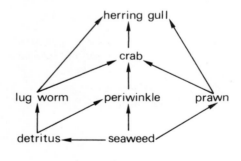

Fig. 18.5

18.2

(a) *Define the following terms used in ecological studies: decomposer; edaphic factor;
energy cycle; microclimate; primary consumer; succession.*

(b) *For a named area where you have personally carried out ecological studies describe how you have:*
   1. estimated the density of any one population:
   2. assessed the variety of organisms present;
   3. established the relationships between various organisms present.

[OLE]

(a) A decomposer is an organism, usually a saprophytic fungus or bacterium, which helps to break down dead and decaying organic materials into simple nutrients. These nutrients can then be recycled through the ecosystem.

An edaphic factor is a soil factor which influences the environment in which organisms live. It may include soil pH, organic content, water content or mineral content.

The term 'energy cycle' refers to the transfer of energy through the ecosystem from sun to top carnivore. It is unidirectional and incurs losses at each trophic level.

The microclimate of a habitat accurately describes the climate of the immediate surroundings of an organism. For example, although the air temperature surrounding a stone wall may be 18 °C the temperature of the air inside the wall where a spider is living could be 30 °C.

A primary consumer is a herbivore which feeds directly upon a primary producer and is, in turn, eaten by a secondary consumer.

A succession describes the sequential existence of different communities within the same area over a number of years.

(b) *Rocky Shore*

1. The number of limpets in a given area, say 100 m², can be determined with the use of a quadrat. The area to be sampled is marked out as 10 m by 10 m and a series of 1 m² quadrats is examined for the presence of limpets along a transect drawn across the two diagonals. The mean number of limpets for each quadrat is determined and multiplied by 100 to give an estimate of the density of the population.

2. To assess the variety of organisms present in this 100 m² area which may include rock pools, stones and sandy areas it is necessary to lay out 1-m belt transects down the shore at regular distances. The number of belt transects chosen will reflect the time available for the study and the degree of accuracy required. Using a quadrat and nets or hand forks, sample the area within each quadrat and record the species present. Care must be taken to hand-search thoroughly. This will involve lifting up stones and seaweed and digging in the sand. All specimens should be handled carefully and the habitat left as undisturbed as possible.

3. A list of species present in an area having been obtained, the relationships between them can be recorded in terms of feeding, parasitism and symbiosis. To do this it is necessary to observe the feeding patterns of each of the organisms either directly or by inference. In some cases one organism may be seen feeding on another or, if not, a study of the mouth parts and stomach contents may reveal the nature of the food eaten. The feeding relationships can then be plotted as food chains and as food webs. More specialized associations such as parasitism and symbiosis require much further observation and study.

18.3

(a) *Define the terms ecosystem, trophic level, and food web.* [6]
(b) *Describe the flow of energy through a named ecosystem.* [12]

[UCLES]

(a) See definitions on page 212 and earlier answers.

(b) Grasslands provide a useful ecosystem for studies of energy flow. 95 to 99 per cent of the sun's radiation is reflected or absorbed and lost as heat of evaporation, or by conduction and convection from clouds and the surface of the earth. The energy that remains is absorbed by the photosynthetic pigments of plants, e.g. the chlorophyll of grasses, herbs and trees and used in the production of organic molecules. The rate of storage of this energy is known as the gross primary production. Of this, 20 per cent is used up by the plants in respiration and photorespiration, and the rest is stored within the plant as net primary productivity. The efficiency of transfer from producer to primary consumer is about 10 per cent and between consumer and consumer about 20 per cent maximum. As energy passes from, say, grass to fieldmice some of the food remains undigested and energy is lost in egestion. This is where much of the net primary productivity is lost. Consumers lose a considerable amount of energy in respiration, excretion and egestion but the energy remaining is available to the organism for growth, repair and reproduction and much of this energy is available to the next trophic level, e.g. owls. Energy which is lost in egestion and excretion is not lost from the ecosystem as it is utilized by decomposers.

18.4

The following data relate to two species of leaf-mining moths. The larvae of one species mine in the leaves of birch trees and the larvae of the other species mine the leaves of oak trees. The data were obtained during an investigation of mortality factors affecting the two species of moths.

Stage in life history	No. of individuals surviving to the end of each stage as a percentage of the number entering the stage.	
	Birch leaf miner	*Oak leaf miner*
Instar 1	92	95
2	95	90
3	82	73
4	65	70
5	45	64
Prepupa	66	71
Pupa	80	64

(a) (i) *If 100 eggs were hatched, calculate the actual number of birch leaf miners which could be expected to survive to the end of the 5th instar stage. Clearly show your method of calculation and correct to the nearest whole number at each stage of your calculation.*

1st instar $\dfrac{92}{100} \times \dfrac{100}{1} = 92$

2nd instar $\dfrac{95}{100} \times \dfrac{92}{1} = 87$

3rd instar $\dfrac{82}{100} \times \dfrac{87}{1} = 71$

$$
\begin{array}{lll}
\text{4th instar} & \dfrac{65}{100} \times \dfrac{71}{1} & = 46 \\[3mm]
\text{5th instar} & \dfrac{45}{100} \times 46 & = 21
\end{array}
$$

(ii) *Given that the survival rate of the comparable 5th instar of the oak leaf miner is 28 per cent, comment on the survival rates of the two species.*

Oak leaf miner survives better overall. There are differences though in survival rate at different times.

[6]

(b) *State two ways in which the survival pattern to the end of the pupal stage of the oak leaf miner compares with that of the birch leaf miner.*

(i) there is a greater mortality for birch miner in prepupal stage

(ii) there is a greater mortality for oak miner in pupal stage

[2]

(c) *Suggest four factors which could limit the population of leaf mining species during their life cycles.*

(i) parasites

(ii) adverse climatic conditions

(iii) increase in numbers of predators

(iv) loss of leaves in autumn

[4]
[L]

18.5

Figure 18.6 is a generalized graph of population growth.

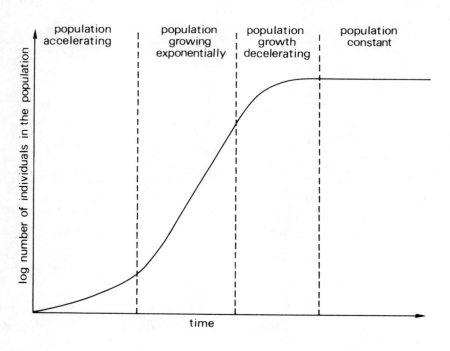

Fig. 18.6

(a) *Why does the initial accelerating phase start slowly?*

    there are few individuals capable of reproducing

(b) *List three environmental factors other than shortage of food, oxygen or water which may cause the growth of the population to decelerate.*

1.   *increase in numbers of predators*

2.   *competition for space*

3.   *increased incidence of disease*

(c) *How do you explain the final phase in which the number of individuals in the population remains constant?*

    birth and immigration rates balance death and emigration rates and no other factors are limiting.

[6]

[UCLES]

18.6

The capture–recapture method is used widely to estimate the size of populations. It involves collection of a sample (number = S_1) of the organism (e.g. an insect) and marking each individual. These are then released, and a second sample collected (number = S_2). The number of marked insects in this second sample is then counted (S_3). The total of insects in the population is then calculated using the formula:

$$N = \frac{S_1 \times S_2}{S_3}$$

(a) *In one such capture–recapture exercise to estimate the size of a population of dragon flies on a stretch of river the first collection resulted in the marking of 250 individuals. Two days later in a second sample of 250, 50 marked individuals were found. Estimate the population size.*

$$\text{Since } N = \frac{S_1 \times S_2}{S_3}$$

$$= \frac{250 \times 250}{50}$$

$$= 1250$$

Population size $= 1250$.

(b) *What assumptions have to be made and what precautions should be taken in using this method?*

1. *Organisms mix randomly within the population.*

2. *Sufficient time must elapse between capture and recapture.*

3. *Organisms disperse evenly within the geographical area of the population.*

4. *It is only applicable to populations restricted geographically.*

5. *Changes in population size are negligible.*

(c) How would you carry out a line transect? What decisions need to be made before using a transect to provide a survey of an area?

Run a tape or string along the ground in a straight line between two poles to indicate the position of the transect. Sample only species actually touching the line or directly beneath it.

1. All points on the line are accessible for examination.
2. It is supposed that there is a transition in habitats and populations along the line.
3. Will this method provide the accuracy required?
4. Is the transect method appropriate for the organisms present?
5. Is this method appropriate in view of the time available?

[7]

[NISEC]

Index

Galapagos Islands 203
gametes 156
gametogenesis 158, 176
gametophyte 156
gaseous exchange surface 72, 78
gastric juice 66
gastrulation 171
Gause 215
gene 186
    flow 203
    isolation 203
    mapping 189
    mutation 178, 179
    pool 202
generative nucleus 156
generator potential 117
genotype 186, 190
geographical distribution 200
geotropism 101
germinal epithelium 158
germination 168, 169
gestation 159
gibberellic acid 106
gibberellins 102, 104, 110, 112, 169
gill 73, 78, 93
    plates 73
    slits 73
gizzard 62
glandular tissue 35
glomerular filtrate 146, 153
glomerulus 146, 150
glucagon 138, 143
glucose 23
glycerol 58
glycogen 15, 22, 64
glycogenesis 135, 138
glycogenolysis 135, 138
glycolysis 71, 74
Golgi apparatus 31
Graafian follicle 158
grafting 156
grana 45, 48
granulocytes 85
graph 11
grey matter 115
gross primary production 214
growth 168, 169
growth hormone 108
growth regulator substance 102
guanine 24
guard cell 33, 45
guttation 80

habitat 212
haemocoel 83
haemoglobin 18, 85
haemolymph 149
haemophilia 196
haemopoietic tissue 35
halophyte 149
Hardy-Weinberg equilibrium 198, 210
Hatch–Slack pathway 47
haustoria 63
Haversian canal 35
hearing 11
heart 93
heartwood 170
heavy-metal tolerance 208
Helix aspersa 61
hepatocytes 138
herbivore 57
hermaphrodite 155, 156
heterodont dentition 60
heterogametic 189, 190
heterosomes 189
heterotroph 44, 57
heterozygous 186
hexose 14, 46

hibernation 137
histology 30
holophytic 44
homeostasis 135, 139, 140, 142
homogametic 189
homologous 177, 201
homology 198, 204
homozygous 186
hormone 113, 118
horse 205
human chorionic gonadotrophin 159
hunger centre 60
hyaline cartilage 127
hybrid 203
hybrid vigour 178, 203
Hydra 61
hydrolysis 13
hydrolytic enzymes 31
hydrophobic 15
hydrophyte 149
hydrostatic skeleton 127
hydrotropism 101
hyperpolarization 115
hypertonic 80, 144, 148
hyphae 62
hypogeal 169
hypopharynx 62
hypothalamus 115, 126, 136, 143
hypotonic 88, 144, 148

ileum 58
immune system 86
inbreeding 202
incipient plasmolysis 81
incisor 60
incomplete dominance 189
indole acetic acid 102, 106, 108
induced-fit theory 21
industrial melanism 203, 204
ingestion 57
inspiration 73
inspiratory centre 73
insulation 137
insulin 18, 108, 138, 143
integuments 157
interfascicular cambium 170
intergranal lamellae 45
interoceptor 116
interphase 175
interstitial cells 158
intraspecific speciation 203
involuntary muscle 36, 128
iodopsin 117
irreversible inhibition 21
irritability 113
islets of Langerhans 138, 143
iso-electric point 13
isolating mechanism 203
isomerism 13, 14
isometric growth 169
isotonic 80, 144

joints 127, 128
juvenile hormone 174

keratin 18
kidney 145, 150
Krebs' cycle 31, 71
Kuppfer cells 139

labium 61
labrum 62
lactation, 159
lacteal 58
lactic acid 72, 129
lacunae 35
lamellae 35, 73
large intestine 58

larynx 73
leaf structure 45
lenticels 170
leucocytes 83, 85
leucoplasts 32
ligament 128
light reaction 44, 45, 49
lignification
    pitted 34
    reticulate 34
    scalariform 34
limiting factor 44, 46, 53, 216
linkage 186, 188
linkage group 188
Linnaeus 200
lipase 64
lipids 15, 64
liver 138, 140
loading tension 85
lock and key theory 21
locus 186
long-day plants 105
loop of Henlé 139, 147, 150
Lumbricus terrestris 62
luteinizing hormone 158
lymph 85
lymph nodes 85
lymphocytes 85
lysosomes 31

macromolecule 13, 14
macronutrient 47, 212
macrophages 35
macrophagous 61
macula 118
magnesium 56
malpighian tubule 149
maltase 27
Malthus 199
maltose 23
mammary glands 159
mandible 61
mass flow 80, 83
mast cells 35
masticate 62
matrix 31, 72
maxillae 61
maximum sustainable yield (MSY) 216
mechanoreceptor 113
medulla 73, 136
medulla oblongata 84, 94, 115, 122
meiosis 175, 176
Mendel 186, 195
menstrual cycle 158
meristem 110
meristematic cells 169
mesophyll 11, 33, 45, 47
mesophyte 149
messenger RNA 20
metabolism 70, 144
metal tolerance 203
metamorphosis 168, 174
metaphase 176
metaphloem 170
metaxylem 34, 170
microfilament 176
micronutrient 47, 212
microphagous 60
micropyle 157, 161, 163
microtubule 32, 176
microvilli 58, 147
middle lamella 32
mitochondria 31, 37, 72, 76
mitosis 175, 180
mitral valve 95
mnemonic 3
molar 60
monoecious 156